THE BLOOD AND ITS THIRD ELEMENT

THE BLOOD
AND ITS
THIRD ELEMENT

ANTOINE BECHAMP

A DISTANT MIRROR

Published by A DISTANT MIRROR

This edition copyright © 2017

ISBN 978-1541159358

All rights reserved, including the right to reproduce this book, or portions thereof, in any form. No part of this book may be reproduced or transmitted in any form or by any means, graphic, electronic, or mechanical, including photocopying, recording, taping or by any information storage or retrieval system, without permission in writing from
the publisher.

The cover image of blood was taken with a Biomedx fiberoptic darkfield microscope, courtesy of Biomedx,
purveyors of microscope and biological terrain analysis systems and related professional training programs.
They are at *www.biomedx.com*.

Web adistantmirror.com.au
Email info@adistantmirror.com.au

CONTENTS

- **5** Editor's Preface
- **8** Translator's Preface
- **13** Author's Preface / 1
- **41** Author's Preface / 2
- **57** Introductory and Historical
- **68** **Chapter One**
  On the nature of fibrin isolated from the clot or obtained by whipping the blood. The blood fibrin. Fibrinous microzymas. Fibrin and oxygenated water. The ferment of fibrin.
- **92** **Chapter Two**
  On the actual specific individuality of the albuminoid proximate principles. The albuminoids. Coagulation. The albuminoids of the fibrin. The albuminoids of the serum. Haemoglobin. Haemoglobin and oxygenated water.
- **111** **Chapter Three**
  The state of the fibrin in the blood at the moment of venesection. The fibrin without microzymas. The haematic microzymian molecular granulations.
- **124** **Chapter Four**
  The real structure of the red blood globule. The microzymas of the blood globules. The blood globules in general.
- **132** **Chapter Five**
  The real nature of the blood at the moment of bleeding. The living parts of the blood protoplasm. The unchangeable character of mixtures of proximate principles. The vitellin microzymas and the blood globules. The vascular system.
- **142** **Chapter Six**
  The real chemical, anatomical and physiological meaning of the coagulation of the shed blood. Coagulation of the blood. The blood of the horse. The serum of the blood. Coagulation of blood diluted with water. Second phase of the spontaneous alteration of the blood in calcined air. Oxygen has no share in the destruction of the globules in the defibrinated blood. Spontaneous alteration of flesh. Spontaneous alteration of milk. Fermentation of the egg. Spontaneous destruction of the cellule of yeast. Spontaneous destruction of tissues. Spontaneous alteration of the blood.
- **172** **Chapter Seven**
  The blood is a flowing tissue and therefore spontaneously alterable. Pasteur and the germs of the air. Robin and the alteration of the blood. Microzymas and spores of schizomycetes. Microzymas and micrococcus. The microzymas and the circulatory system. Comparison of the microzymas of the blood, the circulatory system, and other tissues. Autonomy of the microzymas.
- **185** **Chapter Eight**
  The microzymas and bacteriology. Ovular and vitellin microzymas. Microzymas and molecular granulations. Geological microzymas. Biological characteristics of microzymas. Microzymas and their perennity. Microzymas and pathology. Phagocytosis. Microzymas and anthrax. Microzymas and disease. Microzymas and microbes. Microzymas and the individual coefficient. Microzymas, life and death. Microzymas, blood and protoplasm. Conclusions.
- **216** Postface

EDITOR'S PREFACE

This book is the last work by Professor Antoine Béchamp, a man who should, by rights, be regarded today as one of the founders of modern medicine and biology. History, however, is written by the winners, and too often has little to do with the truth. The career of Antoine Béchamp, and the manner in which both he and his work have been written out of history, are evidence of this.

During his long career as an academic and researcher in nineteenth century France, Béchamp was widely known and respected as both a teacher and a researcher. As a leading academic, his work was well documented in scientific circles. Few made as much use of this fact as Louis Pasteur, who based much of his work on plagiarising and distorting Béchamp's research. In doing so, Pasteur secured for himself an undeserved place in the history of medical science.

There have been several excellent books written, mainly in the early decades of the twentieth century, which explain in detail the plagiarisms and injustices which Pasteur and his allies inflicted on Béchamp. Among these are *Pasteur Exposed* (previously published as *Béchamp or Pasteur?*) by Ethel Douglas Hume, and *The Dream and Lie of Louis Pasteur* (previously *Pasteur, Plagiarist, Imposter*) by R. Pearson.

The Blood and its Third Element is Béchamp's explanation of his position, and his defence of it against Pasteur's mischief. It was his last major work, and as such it embodies the culmination of his life's researches.

This book contains, in detail, the elements of the microzymian theory of the organization of living organisms and organic materials. It has immediate and far reaching relevance to the fields of immunology, bacteriology, and cellular biology; and it shows that more than 100 years ago, the germ, or microbian, theory of disease was demonstrated by Béchamp to be without foundation.

The reader should be aware when reading *The Blood and its Third Element* that in formulating his microzymian theory of biological organisation, Béchamp in no way sought to establish it as the last word on the subjects of disease, its transmission, general physiology, or indeed the organisation of living matter itself. Béchamp worked continuously until a few weeks before his death; and if he were working now, he would no doubt still regard his work as unfinished, and subject to revision and development.

It is no accident, but rather a vindication of Béchamp's theories, that many researchers over the course of the twentieth century and up to the present have arrived at conclusions in various disciplines that support the microzymian model.

In the United States during the 1920s and '30s, Royal Rife's microscope revealed processes of life which confused many of Rife's contemporaries, but which would have made perfect sense to Béchamp. The medical establishment, however, was disturbed by the implications of Rife's discoveries, especially so when he began curing diseases, including cancer, with electromagnetic frequencies. Rife and his discoveries were soon consigned to that special anonymity which is reserved for those who threaten the status quo. To maintain the profits of the drug companies and the authority of the medical establishment, no expense or effort is too great, and by the time Rife died, his work was all but forgotten. The authorities confiscated and destroyed all of his equipment and writing that they could get their hands on. Fortunately, in recent years, interest in his work has revived, as a search on the internet will demonstrate.

Contemporary researchers whose work connects with that of Béchamp include Gaston Naessens (*www.cerbe.com*), whose 'somatids' are without doubt what Béchamp described as 'microzymas'. Naessens has gone further than Béchamp, though, aided by his revolutionary microscope technology, and has identified the various stages of the somatid life cycle.

Just recently, Dr. Philippa Uwins of the Centre for Microscopy and Microanalysis at the University of Queensland in Australia has been making headlines with her work documenting the existence of 'nanobes', which she describes as involving the "morphological and microstructural characterisation of novel nano-organisms".

One can't help but think that Béchamp, Rife, Naessens and Dr Uwins are all talking about the same thing.

EDITOR'S PREFACE

There is no single cause of disease. The ancients thought this, Béchamp proved it and was written out of history for his trouble. The relevance of his work to the dilemmas that plague modern medical science remains as yet unrealized.

Fortunately, though, there are streams of modern research such as the ones mentioned previously that are heading in the right direction, even though they are encountering resistance and cynicism. This book is being republished in the hope that the information it contains can contribute to that research.

This new edition has been reset, in a new layout that will hopefully make the content more accessible. Wherever it has been possible without altering the intent of the author, archaic or ambiguous use of English has been brought up to date.

The footnotes are either Professor Béchamp's or Montague Leverson's. Where they belong to Leverson, they are enclosed by square brackets.

When the letters C.R. appear in a footnote, they denote the *Comptes Rendus* (trans. *transactions*) of the various French academies cited in the text.

D. L. Major

TRANSLATOR'S PREFACE

On October 16th, 1816, at Bassing, in the department of Bas-Rhein, was born a child by whose name the nineteenth century will come to be known, as are the centuries of Copernicus, Galileo and Newton by their names.

Antoine Béchamp, the babe of 1816, died on the 15th April, 1908, fourteen days after he was first visited by an aged American physician between whom and himself a correspondence had passed for several years on the subject of the researches and wonderful discoveries of Professor Béchamp and his collaborators. The American physician made his visit to Paris for the purpose of becoming personally acquainted with the Professor, who, as his family stated, had looked forward with eager anticipation to such a visit.

The translator had long previously submitted an extensive summary of the professor's physiological and biological discoveries, by whom it was revised and approved.

This was intended to be introduced as a special chapter in an extensive work on inoculations and their relations to pathology, upon which the translator of this work had been engaged, almost exclusively, for some fourteen years.

But in the lengthy and nearly daily interviews between Professor Béchamp and myself, which, as just shown, closely preceded the former's death, I suggested that instead of such summary it would be better to place before the English speaking peoples an exact translation into their language of some, at least, of the more important discoveries of Professor Béchamp—especially as, in my opinion, it would not be easy to carry out among them the conspiracy of silence by means of which his discoveries had been buried in favour of distorted plagiarisms of his labours which had been productive of such abortions as the microbian or germ theory of disease, "the greatest scientific silliness of the age," as it has been correctly styled by Professor Béchamp.

To this suggestion Professor Béchamp gave hearty assent, and told me to proceed exactly as I might think best for the promulgation of the great truths of biology, physiology, and pathology discovered by

TRANSLATOR'S PREFACE

him, and authorised me to publish freely either summaries or translations into English, as I might deem most advisable.

As a result of this authorisation, the present volume is published, and is intended to introduce to peoples of the English tongue the last of the great discoveries of Professor Béchamp.

The subject of the work is described by its title, but it is well to remind the medical world and to inform the lay public that the problem of the coagulation of the blood, so beautifully solved in this volume, has until now been an enigma and opprobrium to biologists, physiologists and pathologists.

The professor was in his 85th year at the time of the publication of the work here translated. To the best of the translator's knowledge it has not yet been plagiarised, and is the only one of the Professor's more important discoveries which has not been so treated; but at the date of its publication the arch plagiarist (Pasteur) was dead, though his evil work still lives.

One of the discoveries of Béchamp was the formation of urea by the oxidation of albuminoid matters.[1] The fact, novel at the time, was hotly disputed, but is now definitely settled in accordance with Béchamp's view. His memoir described in detail the experimental demonstration of a physiological hypothesis of the origin of the urea of the organism, which had previously been supposed to proceed from the destruction of nitrogenous matters.

By a long series of exact experiments, he demonstrated clearly the specificity of the albuminoid matters and he fractionised into numerous defined species albuminoid matters which had until then been described as constituting a single definite compound.

He introduced new yet simple processes of experimentation of great value, which enabled him to publish a list of definite compounds and to isolate a series of soluble ferments to which he gave the name of *zymases*. To obscure his discoveries, the name of *diastases* has often been given to these ferments, but that of *zymas* must be restored. He also showed the importance of these soluble products (the zymases) which are secreted by living organisms.

He was thus led to the study of fermentations. Contrary to the then generally accepted chemical theory, he demonstrated that the alcoholic fermentation of beer yeast was of the same order as the phenomena which characterise the regular performance of an act of animal life—digestion.

In 1856, he showed that moulds [2] transformed cane sugar into invert sugar (glucose) in the same manner as does the inverting ferment secreted by beer yeast. The development of these moulds is aided by certain salts, impeded by others, but without moulds there is no transformation.

He showed that a sugar solution treated with precipitated calcic carbonate does not undergo inversion when care is taken to prevent the access to it of external germs, whose presence in the air was originally demonstrated by him.[3] If to such a solution the calcareous rock of Mendon or Sens be added instead of pure calcic carbonate, moulds appear and the inversion takes place.[4]

These moulds, under the microscope, are seen to be formed by a collection of molecular granulations which Béchamp named *microzymas*. Not found in pure calcic carbonate, they are found in geological calcareous strata, and Béchamp established that they were living beings capable of inverting sugar, and some of them to make it ferment. He also showed that these granulations under certain conditions *evolved into bacteria*.

To enable these discoveries to be appropriated by another, the name *microbe* was later applied to them, and this term is better known than that of microzyma; but the latter name must be restored, and the word microbe must be erased from the language of science into which it has introduced an overwhelming confusion. It is also an etymological solecism.[5]

Béchamp denied spontaneous generation, while Pasteur continued to believe it. Later he, too, denied spontaneous generation, but he did not understand his own experiments, and they are of no value against the arguments of the sponteparist Pouchet, which could be answered only by the microzymian theory. So, too, Pasteur never understood either the process of digestion nor that of fermentation, both of which processes were explained by Béchamp, and by a curious imbroglio (was it intentional?) both of these discoveries have been ascribed to Pasteur.

That Lister did, as he said, most probably derive his knowledge of antisepsis (which Béchamp had discovered) from Pasteur is rendered probable by the following peculiar facts.

In the earlier antiseptic operations of Lister, the patients died in great numbers, so that it came to be a gruesome sort of medical joke to say that "the operation was successful, but the patient died." But Lister was a surgeon of great skill and observation, and he gradually reduced his employment of antiseptic material to the necessary and not too large dose, so that his operations "were successful and his patients lived."

Had he learned his technique from the discoverer of antisepsis, Béchamp, he would have saved his earlier patients; but deriving it second hand from a savant (sic) who did not understand the principle he was plagiarising,[6] Lister had to acquire his subsequent knowledge of the proper technique through his practice, i.e. at the cost of his earlier patients.

Béchamp carried further the aphorism of Virchow—*Omnis cellula e cellula*—which the state of microscopical art and science at that time had not enabled the latter to achieve. Not the cellule but the microzyma must, thanks to Béchamp's discoveries, be today regarded as the unit of life, for the cellules are themselves transient and are built up by the microzymas, which, physiologically, are imperishable, as he has clearly demonstrated.

Béchamp studied the diseases of the silk worm then (1866) ravaging the southern provinces of France and soon discovered that there were two of them—one, the pébrine, which is due to a parasite;[7] the other, the flacherie, which is constitutional.

A month later, Pasteur, in a report to the Academy of his first silkworm campaign, *denied* the parasite, saying of Béchamp's observation, "that is an error." Yet in his second report, he adopted it, as though it were his own discovery!

The foregoing is but a very imperfect list of the labours and discoveries of Antoine Béchamp, of which the work here translated was the crowning glory.

The present work describes the latest of all the admirable biological discoveries of the Professor Béchamp. It is proposed to follow it up with a translation of *The Theory of the Microzymas and the Microbian System* now in course of translation; and *The Microzymas,* the translation whereof is completed. Other works will, it is hoped, follow, viz.: *The*

Great Medical Problems, the first part of which is ready for the printer, *Vinous Fermentation,* translation complete; and *New Researches upon the Albuminoids,* also complete.

The study of these and of the other discoveries of Professor Béchamp will produce a new departure and a sound basis for the sciences of biology, of physiology and of pathology, today floating in chaotic uncertainty and confusion; and will, it is hoped, bring the medical profession back to the right path of investigation and of practice from which it has been led astray into the microbian theory of disease, which, as before mentioned, was declared by Béchamp to be the "greatest scientific silliness of the age."

Montague R. Leverson
London, 1911

NOTES

A few notes have been appended by the translator; these are distinguished from the author's by being enclosed in square brackets—[ ] and by the phrase —*Trans.* which appears at the end of the note.

The letters *C.R.* are used for the words *Comptes Rendus* (trans. *transactions*) of the various French academies cited in the text.

AUTHOR'S PREFACE / 1

> There is nothing but what ought to be.
> —*Galileo*
>
> Nothing is created, nothing is lost.
> —*Lavoisier*
>
> Nothing is the prey of death:
> all things are the prey of life.
> —*The author*

An historian of the founders of modern astronomy recently related that the philosopher Cleanthus, three millennia before our era, wished to prosecute Aristarchus for blasphemy, for having believed that the earth moved, and having dared to say that the sun was the immovable centre of the universe. Two thousand years later, human reason having remained stationary, the wish of Cleanthus was realized. Galileo was accused of blasphemy and impiety for having, like Copernicus and following Aristarchus, maintained the same truth; a tribunal condemned his writings, and forced him to a recantation which his conscience denied.

The following is the judgement of the historian upon this event:

> " Never perhaps has the generous detestation of the public conscience for intolerance shone forth more strongly than around the name of Galileo.
>
> The narrative of his misfortunes, exaggerated like a holy legend, has affirmed, while avenging him, the triumph of the truths for which he suffered; the scandal of his condemnation will forever vex in their pride those who would oppose force to reason; and the righteous severity of opinion will preserve its inconvenient remembrance as an eternal reproach thrown in their teeth to confound them."

The "righteous severity of the judgement" which preserves the inconvenient memory of the sufferings of Galileo, it is well to mention, is that of the scholarly and learned members of Academies whereof the author forms part. It is agreed; yes, intolerance is odious and hateful, the situation of Galileo was particularly horrible. He was forced to go to church and pronounce with a loud voice the abjuration dictated to him.

> " I, Galileo, in the seventieth year of my age, on my knees before your Eminences, having before my eyes the holy gospels, which I touch with my own hands, I abjure, I curse, I detest, the error and the heresy of the movement of the earth."

There is no more atrocious torture than this brutal violence against the conscience of a man. It is the greatest abuse of force and pride when we know that it was the priests of Jesus Christ who perpetrated it.

The theologians of the holy office were not competent to judge the astronomer Galileo, yet they in their ignorance undertook to proscribe an opinion which differed from their own as being erroneous and contrary to the holy Scriptures, which, said the Popes, "were dictated by the mouth of God himself." In truth what did they know about it? Assuredly it is distressing to observe how long human reason can remain at the same point.

It is then interesting to know whether the lesson taught by the condemnation of Galileo has been properly learned, and if three centuries later "the righteous severity of the judgement against those who would still resist the power of reason" would be able to protect those who labour disinterestedly for the triumph of the truth. Have those who, for the large public, are the authoritative judges of the value of the discoveries of others become less intolerant, or at least more impartial, less prompt to pronounce against opinions which they do not share, and less anxious to deny facts than to test them?

And if the lesson has not been learned, it is not less interesting to ask whether it is "human reason" which must be held responsible; if it might not instead be "pettifogging" ratiocination, the abuse of reasoning warped by passion and too often by personal interest which overcomes private conscience and leads the public astray.

The history of a discussion wherein chemistry and physiology closely united were interested, which agitated the second half of the 19th century, is well adapted to show that human nature has not changed since the time of Cleanthus, and that there always exist people ready to associate themselves together to contradict or insult the unfortunate wretch who has devised some new theory, based upon unsuspected facts, which would compel them to reform their arguments and abandon their prejudices.

This work upon the blood, which I present at last to the learned public, is the crown to a collection of works upon ferments and fermentation, spontaneous generation, albuminoid substances, organization, physiology and general pathology which I have pursued without relaxation since 1854, at the same time with other researches of pure chemistry more or less directly related to them, and, it must be added, in the midst of a thousand difficulties raised up by relentless opponents from all sides, especially whence I least expected them.

To solve some very delicate problems I had to create new methods of research and of physiological, chemical and anatomical analysis. Ever since 1857 these researches have been directed by a precise design to a determined end: the enunciation of a new doctrine regarding *organization and life*.

It led to the microzymian theory of the living organization, which has led to the discovery of the true nature of blood by that of its third anatomical element, and, at last, to a rational, natural explanation of the phenomenon called its spontaneous coagulation.

But the microzymian theory, which is to biology what the Lavoisierian theory of matter is to chemistry, and which is founded on the discovery of the microzymas, living organisms of an unsuspected category, has been attacked in its principle, by denying the very existence of the microzymas.

Since this was so, if the assertion that the microzymian theory of the living organization gives to biology a base as solid as does the Lavoisierian theory to chemistry be deemed imprudent, well, I choose to commit this imprudence, and to be imprudent to the end, and to struggle against a current of opinion which is the more violent, as will be seen, the more it is artificial.

It was the boldest of those who deny the fact of the existence of the microzymas who wrote:

> "Whenever it can be done, it is useful to point out the connection of new facts with earlier facts of the same order. Nothing is more satisfying to the mind than to be able to follow a discovery from its origin to its latest development."[1]

That is very well and fine, the more so that the author took good care not to follow this wise precept; let us ascend then to the sources.

Two centuries after Galileo, we were still in the Aristotelian hypothesis regarding matter, but reinforced by the alchemical hypothesis of *transmutation* and the Stahlian one of *phlogiston*. It was readily conceded that matter could *of itself* become living matter, animated, such as it is in plants and animals; thus it was that spontaneous generation was still generally accepted.

Charles Bonnet himself said that organization was *the most excellent modification of matter*; nevertheless that learned naturalist and philosopher attempted to oppose spontaneous generation by imagining in turn the hypothesis of encapsulation and that of pre-existing germs universally diffused, whereof Spallanzani made use to refute the experiments and conclusions of the sponteparist Needham.

On the other hand, to sustain Needham, Buffon invented the hypothesis of *organic molecules,* not less universally diffused, whose substance, distinct from common matter, called raw matter, helped to explain the growth of plants and animals, as well as spontaneous generation.[2]

Fermentations and ferments were very simply explained. Macquer, in 1772, regarded it as certain that vegetable and animal matters, abstracted from living organisms, under certain conditions of the presence of water and of contact, at least momentarily, with the air and of temperature, become altered of themselves, and ferment, becoming putrid in producing the ferment.

And according to the same principles it was said that water could transmute itself into earth, the earth into a poplar, and that the blood begets itself by the transmutation of flesh into the flowing liquor.

Such in a few words was the condition of science upon these questions before the advent of Lavoisier. In the Lavoisierian theory there is no matter other than that of simple bodies, which are heavy, indestructible by the means at our disposal, and always reappearing the same, not withstanding all the vicissitudes of their various combinations among themselves and the changes of states or allotropic modifications they might undergo. No transmutations and no phlogistication to explain the phenomena.

In this theory, matter is only mineral, simple bodies being essentially mineral. There is no living or animal matter, no matter essentially organic.

That which, long after the time of Lavoisier, chemists have called organic matters are only innumerable combinations in the various proportions which carbon, hydrogen, oxygen, and nitrogen can form, often with other simple bodies at the same time—sulphur, phosphorus, iron, etc, carbon being always present, so that what is called organic matter in modern chemistry is only various combinations of carbon with the simple bodies mentioned.

In fact, Lavoisier, after his demonstration that water did not become transmuted into earth, nor earth into plants, asserted clearly that plants draw their food from the air, as was verified later. He even asserted that animals obtained the materials for their nutrition from plants, thus demonstrating that plants effected the synthesis of the substance without which animals could not exist. Even respiration was only a common phenomenon of oxidation.

The substance of plants and animals being only combinations of carbon with hydrogen and oxygen, with the addition of nitrogen for animals, it is very interesting to recall shortly what Lavoisier thought of the putrefaction of these substances and of fermentation.

Like everybody, he knew that the juice of grapes or apples enters into fermentation of itself to produce wine or cider, and he wrote the following equation:

$$grape = must = carbonic\ acid + alcohol$$

To demonstrate this, he reduced the experiment to the employment of sugar, which he called a vegetable oxide, and of water and a ferment. The following is his account of the experiment:

> "To ferment sugar, it must first be dissolved in about four parts of water. But water and sugar, no matter what proportions be employed, will not ferment alone, and equilibrium will persist between the principles

(the simple bodies) of this combination if it is not broken by some means.

A little yeast is sufficient to produce this effect and to give the first movement to the fermentation; it then continues of itself to the end. The effects of vinous fermentation reduced themselves to separating the sugar into two portions, to oxygenize the one at the expense of the other to produce carbonic acid of it; to deoxygenize the other in favour of the former to make alcohol of it; so that if it were possible to recombine the alcohol and carbonic acid, the sugar would be reformed."

It is thus clear that Lavoisier instead of the equation regarding the *must* might have written thus:

$$sugar \; = \; carbonic\; acid \; + \; alcohol$$

Lavoisier intended to give elsewhere an account of the effects of yeast and of ferments in general, which he was prevented from doing. But it can be seen from his *Treatise upon Elementary Chemistry,* published in 1788, that he had established that yeast is a quarternary nitrogenised body, and that that which remained of it at the end of the fermentation contained less nitrogen, and that besides the alcohol, a little acetic acid was formed. Lavoisier also found that after distillation there remained a fixed residue representing about 4% of the sugar. We shall see later the importance of these remarks.

It might thereafter have been anticipated that Lavoisier should explain the phenomena of the putrid fermentation of vegetable and animal substances "as operating by virtue of very complicated affinities" between the constituted principles of these substances (the simple bodies), which in this operation cease to be in equilibrium so as to be constituted into other compounds.

Bichat, who died in 1802 at the age of 31, had been much struck by the results of the labours of Lavoisier. He could not accept a living matter constituted of pure chemical compounds whereof the simple elements are the constituent principles. He imagined, then, that the only living things in a living being are the organs composed of the tissues, of which he distinguished twenty-one as elementary anatomical elements, as the elementary bodies are chemical elements. Such was the first influence of the Lavoisierian theory upon physiological anatomy; it was thus that in 1806 in the third edition of his *Philosophie Chimique*, Fourcroy said:

> "Only the tissue of living plants, only their vegetating organs, can form the matters extracted from them, and no instrument of art can imitate the compositions which are prepared in the organized machines of plants."

Let us bear in mind that Bichat had been led by the Lavoisierian theory of matter to lay down a new principle of physiology. As Galileo had laid down the metaphysical principle "nothing is but what ought to be", Dumas drew from the chapter on fermentation of Lavoisier's treatise the following principle, which is also a necessary one: *"nothing is created, nothing is lost."*

We have above rapidly sketched the state of the relations of chemistry and physiology as well as the state of the subject of fermentations at the beginning of the nineteenth century; we will now see what they were at the commencement of the second half of that century, in about 1856.

The chemists, thanks to direct analytical methods which were more and more perfected, had isolated a great number of incomplex compounds, acids, alkaloids, neutral or having diverse functions, from vegetable and animal substances. Those incomplex compounds were more and more exactly specified under the name of *proximate principles* of plants and of animals, nitrogenised ternaries and quarternaries.

Among the nitrogenised proximate principles, a number of them were distinguished as soluble or insoluble, and also uncrystallisable, such as the albumin of the white of egg and of the serum of blood, caseum (later called casein) of milk, the fibrin of the blood and that of the muscles, the gelatine of the bones, the gluten of wheat, the albumin of the juices of plants, etc. In time, the similarity of their composition and of certain of their common properties with the albumin of the white of egg led to these matters being formed into the groups of the albuminoid matters.

Lavoisier knew these albuminoid matters only in so far as they were nitrogenised animal matters.

Now after the discovery of gluten, of vegetable albumen, and nitrogenised quarternaries like beer yeast, it was admitted that they were the ferment of vinous fermentation. Then, generalising, it came to be thought that albumin, the albuminoids in general, became or were directly the ferment, while the ternary proximate principles, such as cane sugar, grape sugar, milk sugar, the other sugars, amylaceous matter, inulin, gum, mannite, etc, were called fermentescible matter.

Matters had reached this point when in about 1836, Cagniard de Latour, resuming the study of beer yeast[3] and of its multiplication during the fermentation which produces beer, regarded it as organized and living, decomposing the sugar into alcohol and carbonic acid by an effect of its vegetation.

That was a conception as original as that of Bichat. It is not because of his having regarded beer yeast as organised and its multiplication during fermentation as a multiplication by vegetation that the conception of Cagniard de Latour is original; it is because he admitted that the fermentation of the sugar operated by an effect of this vegetation, that is to say, owing to a physiological act.

That was an absolutely new point of view; beer yeast, the only isolated ferment known, ceased to be regarded as a precipitate of albuminoid matter which had become insoluble, and was henceforth looked upon as a living being! Consequently yeast ceased to be regarded as the reagent that Lavoisier had said was able to disturb the equilibrium of the simple bodies which constituted sugar.

Also, soon afterwards, Turpin, the botanist, interpreted the effect of the vegetation of Cagniard by saying that the globule of yeast was a cellule which decomposed sugar in nourishing itself. Dumas went further, and asserted that the ferments, the yeast, behaved as do animals when feeding, and that, for the orderly maintenance of the life of the yeast, there was needed, as for animals, nitrogenised albuminoid matter as well as sugar.

In Germany, Schwann supported the opinion of Cagniard de Latour while broadening the question; he supposed that no animal or vegetable substance altered of itself and that every phenomenon of fermentation presupposed a living ferment. To prove this, he experimented as Spallanzani had done—improving upon his method in order to demonstrate that the infusoria or ferments had their origin in the germs of the air. The experiments of Schwann were confirmed by others.

But the conception of Cagniard de Latour did not prevail, nor especially the interpretation of Turpin and Dumas. It was not denied that infusoria or moulds existed in the mixtures in a state of alteration, but it *was* denied that they were the agents of the fermentation; this would begin of itself and the altered matter was regarded as evidence in favour of either spontaneous generation or the production of these living products by the germs of the air.

The discovery of diastase and synapse, soluble and nitrogenised quarternaries like yeast, was held to legitimize the refusal to consider yeast as acting because it was organized and living.

Now because these substances were reagents of rare power for transforming certain fermentescible matters in aqueous solution, the transformations were called fermentation, and these reagents were called ferments; and it was said that it is not because they are organized and living that the ferments act to effect the phenomena of fermentation.

Then the opponents of the doctrine of Cagniard de Latour and Schwann, with regard to fermentations and the relations of chemistry to physiology, triumphed so completely that opinions reverted to the point maintained in 1788. The principle of Bichat's doctrine was lost to view; not only was it proposed that vegetable and animal matters altered of themselves under the conditions specified by Macquer, but so too the proximate principles extracted from them, even cane sugar, the aqueous solution whereof Lavoisier had declared to be unalterable.

In short, the old hypothesis of germs of the air, which Schwann had revived, was completely lost to view.

Nothing is better fitted to convince one that the human soul during the second half of the 19th century has remained the same as it was in the times of Galileo and of the inquisition than to reflect upon the sequel of the history I have just sketched out.[4]

I will now describe the fundamental experiment, the results whereof have completely changed the aspect of science with regard to the relations of chemistry and physiology with fermentation, such as they were still imagined to be at the end of the year 1857, after the theory of Cagniard de Latour in relation to yeast had been rejected.

In 1854, it was conceded that cane sugar dissolved in water altered of itself and became transformed into what is called invert sugar, because the solution which deviated the plane of polarisation to the right before the alteration deviated it to the left afterwards. The inverted sugar was also called grape sugar. The phenomenon of this alteration was called *inversion*.

With reference to other researches I resolved to verify the fact, and in the month of May, 1854, I left to themselves in a closed flask, in the presence of a small volume of air, at ordinary temperature and in a diffused light, some aqueous solutions of pure cane sugar. After several months, I found that the sugar solutions in pure distilled water were partly inverted.

At the beginning of 1855 I published the observation as a verification of the fact, but I mentioned at the same time the presence of a mould in the inverting liquor. It is not an unusual thing to see moulds appear in aqueous solutions of the most diverse substances.

That was why, in the then state of science and given the contradictory assertions regarding the experiments of Schwann, I would not assert anything beyond the fact. I noted merely that in the solutions to which I had added chloride of calcium, or chloride of zinc, the inversion had not taken place and no mould had appeared. To find an explanation of these differences I made various experiments, commencing in 1855 and continuing to the month of December, 1857.

Among these experiments, all accordant with one another, I select two, because, reducing the problem to its simplest expression, they leave no room for doubt concerning the legitimacy of the conclusions I deduced from them.

The first conclusion was that the solution of cane sugar in distilled water remains indefinitely unchanged when, having been boiled, it is preserved in an absolutely full closed vase.

The second was that the same solution, whether boiled or not, left in a closed vessel in the presence of a limited volume of air permits the appearance of colourless moulds, generally myceliennated, and the solution becomes completely inverted in the course of time, while the liquor reddens litmus paper, that is to say, it becomes acid. To prove that the volume of air left in the closed flask has nothing to do with the inversion, it suffices to add beforehand a small quantity of creosote[5] or a trace of sublimate of mercury to ensure that the liquid shall not become acid, or mouldy, and that the sugar will remain unchanged.

These two experiments clearly demonstrated to me that the presence of the air was essential for the inversion to take place and for the moulds to be born, and at the same time that the volume of air left present could not operate the inversion.

It was then necessarily the developed moulds which were the agents of the phenomena observed. But myceliennated moulds are true microscopic plants, and consequently organized and living. I proved that they were nitrogenised and that, introduced into creosoted sugar water, they inverted the cane sugar much more rapidly than during their development. Nevertheless these moulds being insoluble, I asked myself: *how do they do it?* And I supposed that it was by an agent analogous to diastase and also thanks to the acid formed; but I have since demonstrated that it was indeed chiefly by means of a soluble ferment which they contain and which they secrete. And the presence of this soluble ferment, and consequently of an albuminoid matter, explained to me how, being nitrogenised, the moulds, when heated with caustic potash, set free an abundance of ammonia.

But these moulds being nitrogenised could not be born of the cane sugar, which I have proven to be exempt from nitrogen. Besides this sugar there was nothing present but distilled water, the mineral substance of the glass, and no other nitrogen than that of the air left in the closed flask; now (thanks to a little creosote or mercuric chloride) the experiment itself showed that these materials could not unite of themselves, by synthesis, to produce the substance of the moulds. Nothing remained to explain the birth of the organized productions other than the old hypothesis of germs; which allowed me no rest until I had discovered their origin and nature.

While waiting to specify them, I admitted that under the conditions of the experiment "germs brought by the air found in the sugared solution a favourable medium for their development";[6] a development during which the new organism, making use of the materials present, effects the synthesis of the nitrogenised and non-nitrogenised materials of its substance.

Under the conditions of the experiment such as I have reported, where there are no other mineral matters than those of the glass, the crop of organized production is necessarily very small, and the inversion as well as the transformations which follow it are very slow.

The addition of certain salts or of creosote hinders the inversion by preventing the development of the germs, either by rendering the medium sterile or by acting directly upon the former.

But the addition of certain other purely mineral salts, even of arsenious acid, had the effect of increasing the harvest and of singularly hastening the inversion and the other phenomena of fermentation which follow it, for if the reaction is prolonged, the acid of which I have spoken above is found to be acetic acid, with, in certain cases, lactic acid, and alcohol in all cases; but to determine the production of this last the mould must be allowed to act for several years. It was thus that I was able to establish that the study made in 1857 was really a phenomenon of fermentation, for the manifestation of which it had not been necessary to employ albuminoid matter, but which, on the contrary, was produced from these matters.

In its simplicity, the experiment was of the same order for physiological chemistry as had been the observation of Galileo with regard to the lamp, hung by a long cord, which oscillated slowly before the altar of the cathedral of Pisa. From that oscillation it was learned that it always beat the same measure, that the duration of the oscillation is independent of its amplitude, and Huyghens discovered the law of the pendulum's oscillation by connecting it with the Galilean principle of falling bodies. The consequences which have sprung from the above experiment have not been less fruitful; some day doubtless there will come a genius like that of Huyghens to extend them and increase their fruitfulness; meanwhile the following are some which I have been able to deduce from it, either in 1857 or subsequently while continuing to experiment. The chief and essential facts of the memoir of 1857 are the following.

1) Cane sugar, a proximate principle, in watery solution, is naturally unalterable even in contact with a limited volume of air, when the solution has been previously creosoted.

2) The solution of cane sugar in contact with a limited volume of air permits the appearance of moulds and the sugar is altered, first of all becoming inverted.

3) If the solution has first had creosote added to it, moulds do not appear and the sugar is not altered.

4) The fact that moulds develop in sugared water, in contact with a small limited quantity of air, forms the verification of the hypothesis of atmospheric germs; in no other way can that fact be explained.

5) Developed moulds invert the cane sugar, even when the solution has first been creosoted, i.e. the creosote which hinders the moulds from being born does not prevent them, when born, from acting. Moulds, being insoluble by reason of their being organized, effect the inversion by means of an agent analogous to diastase; that is to say, by means of a soluble ferment.

6) The totality of the phenomena of the non-spontaneous alteration of cane sugar and the production of an acid and of alcohol prove it to be a fermentation both of moulds and of ferments.[7]

These facts, studied more attentively, showed clearly, contrary to what had before been believed, that albuminoid matter was not necessary for the birth of these ferments; and also that the soluble ferments were not the products of the alteration of some albuminoid matter, since the mould produced at once the albuminoid matter and the soluble ferment by virtue of its physiological functions of development and nutrition.

Thus it resulted that the soluble ferment was allied to the insoluble by the relation of product to producer; the soluble ferment being unable to exist without the figured ferment, which is necessarily insoluble.

Further, as the soluble ferment and the albuminoid matter, being nitrogenised, could only be formed by obtaining the nitrogen from the limited volume of air left in the flasks, it was at the same time demonstrated that the free nitrogen of the air could help directly in the synthesis of the nitrogenised substance of plants. Up to that time this had been a disputed question.

Thenceforward it became evident that since the synthesis of the materials of the substance of moulds, of ferments, is necessarily produced by intussusception within the organism of these moulds, it must necessarily be that all the products of fermentation are produced there and that they are secreted therein as was secreted the soluble ferment which inverted the cane sugar.

Hence I became assured that that which is called fermentation is, in reality, the phenomenon of nutrition; i.e. the assimilation, dissimilation, and excretion of the products dissimulated.

Without doubt, these views were in conformity with the conceptions of Cagniard de Latour, even to those of Schwann and to the more precise view of Turpin and especially of Dumas; but in complete disagreement with those of their opponents, Liebig and his followers, some of whom denied that yeast was living, and held it to be nitrogenous matter in a state of decomposition, and others that it acted in so far as it was nourished, by an action of *extalyic contact,* an occult cause, and that it effected the decomposition of sugar in the same manner as did platinum that of oxygenated water.

We must then demonstrate that that which was true of the moulds was so in the same sense as in the case of beer yeast and of the ferment of the lees of wine; that is to say, that the cellules of these ferments invert cane sugar under the same conditions, in spite of the creosote, and *before* any other phenomenon of transformation is produced. It is found, in effect, that the yeast contains the soluble ferment which inverts, as the mould also contains it.

Nevertheless, the opponents of the conception of Cagniard de Latour and Schwann could always object that if the creosote prevents the cane sugar from being altered, it would not be the same in the case of a mixture containing albuminoid matter, and that consequently, if in the mixture of sugared water and beer yeast, the cane sugar was inverted, it was because beer yeast, an albuminoid substance, continued to be altered in spite of the creosote.

I replied by demonstrating that under the same conditions as the cane sugar all the true proximate principles, including soluble and insoluble albuminoids, even the most complex mixtures of proximate principles, remained unchanged, nothing organized appearing in them—provided that in the cases wherein cane sugar is present, the inverting soluble ferment does not exist among these proximate principles, because creosote does not prevent double ferments from reacting.

Two contemporary experiments of that fact greatly impressed me.

The first relates to milk. Everybody except Dumas regarded milk as an emulsion, as a pure mixture of proximate principles. Now, it is known that, like blood, it alters and clots after it is drawn, as Macquer said in the last century (the 18th).

This furnished an opportunity to verify the fact of the unchangeableness of mixtures of proximate principles when creosoted.

The milk of a cow was then creosoted while being drawn, by receiving it into vessels washed with boiling creosoted water divided into three portions; one of which was left with a limited volume of air present; a second was left without any, and in the third the air was expelled by a current of carbonic acid gas. To my very great surprise, the milk altered, became sour and clotted, almost as quickly as if no creosote had been added. And lastly, which surprised me most of all, shortly after the coagulation was completed, there was a crowd of bacteria in every part of the clot.

The second experiment relates to the chalk which chemists employed, as calcic carbonate, in their experiments even upon fermentation, and which, like them, I employed to preserve the neutrality of the media.

One day, some starch made of potato fecula had some chalk added to it to prevent it turning sour and was left in an oven at 4° to 45°C (104° to 113°F). I expected to find the starch with the same consistency as before; on the contrary, it was liquefied. "The germs of the air," I said.

I repeated the experiment, creosoting the boiling starch and added some of the same chalk; again liquefaction! Much astonished, I repeated the experiment, replacing the chalk with pure artificial calcic carbonate; this time the creosoted starch was not liquefied, and I preserved it in this state for ten years.

These two experiments, in their simplicity, were of the same order, equally fundamental as that of the inversion of sugar by moulds, but they embarrassed me much more.

It was not until after other researches and after having varied and controlled them that I placed them before the learned societies of Montpellier (1863) and informed Dumas of them in a letter which he thought fit to publish,[8] in which I stated that some of the calcareous earths and milk contained living beings already developed.

And here are three other experiments, not less fundamental, which verify the first three:

1) I had ascertained that in the fermentation of cane sugar by moulds born of atmospheric germs, in a watery solution of sugar, acetic acid is produced; why is it not also produced in fermentation by beer yeast? And I shall prove that there is, in fact, produced at the same time only a very small quantity of acids homologous to acetic acid.

2) Beer yeast inverting cane sugar as do moulds, I tried to isolate from the yeast the soluble ferment it produces, as one can readily obtain as much beer yeast as may be required. I will say here how I proceeded to isolate it directly. Brewery yeast, pure, washed and drained, was treated with powdered cane sugar in suitable quantity. The mixture of the two bodies became liquefied and the sugar was entirely dissolved. The product of the liquefaction was thrown upon a filter. If the operation is performed on a sufficiently large quantity, there results the flowing off of an abundant limpid liquid before any indication of fermentation is manifested.

   The filtered liquid, being treated with alcohol, furnishes (as does an infusion of sprouted barley to precipitate its diastase) a rather considerable white precipitate, whereof the part soluble in water is the required soluble ferment. There could be no further doubt; this soluble ferment forms part of the very substance of the content of the cellule of the yeast. I gave it the name first of *zymas*, and later that of *zythozymas*.

3) The cellule of yeast, being a living organism, ought, being insoluble, to possess a vital resistance and should permit only such things to issue from its being as were disassimilated in it.

   Now, in effect, pure yeast, subjected to a methodical washing with distilled water, yields to it at first scarcely anything, only a trace of zythozymas and phosphoric acid. But there comes a time when it yields enormously, then less and less, until it has lost nearly 92% of its substance, preserving its form with its tegument distended with water.

   The observation suggested a comparison with the famous experiment of Chossat upon starving dogs. To compel the yeast to dwell in pure water would be to deprive it of nourishment; to submit it to a regimen of starvation would force it to devour itself. Pure yeast, steeped in creosoted distilled water, absolutely protected from air, disengages pure carbonic acid for a long time, producing alcohol, acetic acid, etc, and at the same time other compounds which it does not make when nourished upon sugar. It exhausts itself thus enormously, remains whole a long time, its tegument preserving its form and, having eliminated its content almost wholly, inverts cane sugar to the end. I thus demonstrated that notwithstanding the creosote, the yeast alters of itself, as does the milk.

The spontaneous alteration of milk and that of yeast seemed to me indisputable proof that neither milk nor yeast was a mixture of proximate principles, but that both of them contain, inherently, the living organized agent which is the cause of their spontaneous alteration, or that consequently, if the chalk liquefies fecula starch, it is because it contains that which can produce the necessary soluble ferment.

It was the experiment of starving the yeast which enabled me to complete the demonstration that the phenomenon called the fermentation of cane sugar by yeast was the digestion of the sugar by the zymas, the absorption of the digested (invert) sugar *by* the cellule, the decomposition of this sugar *in* the cellule being the result of the complex phenomenon of *assimilation,* followed necessarily by disassimilation and of elimination. The products eliminated were carbonic acid, alcohol, acetic acid, etc, the same as with man the products of disassimilation—urea, etc.—come from man and reunite in part in urine.

While I was thus experimenting to develop the consequences of the memoir of 1857 and discovered the *zythozymas* in the yeast, I also discovered *anthozymas* in flowers, *morozymas* in the white mulberry, and the *nefrozymas* of the kidneys in the urine as a product of the function of the kidneys, in order to demonstrate that as the moulds form and secrete their soluble ferment, plants and animals form theirs in their organs, and I shall demonstrate besides that the leucocytes of pus even produce a zymas in the pus.

The phenomenon called fermentation is then the phenomenon

of nutrition, which is being accomplished in the ferment, in the cellule of the yeast, in the same manner as the phenomenon of nutrition is accomplished in the animal, and following the same mechanism by the same means. This was the fundamental idea of my memoir *Upon Fermentations by Organized Ferments* which dates from 1864.[9]

I will revert later, with details, to this work, which is fundamental. I mention it now only as a verification of the conception of Dumas of which mention has before been made; it was in that work that for the first time the word *zymas* is employed to designate the soluble ferment which yeast contains performed, distinguishing the soluble ferments as agents of a different order from the figured ferments and effecting transformations also of a different order.

For the history one should read, in the *Jahresbericht* of Heinrich Will for 1864, how this was regarded as new in Germany and was favourably appreciated.

It is difficult, however, to realize the resistance which was offered from many sources to the demonstration that the phenomenon of fermentation is a phenomenon of nutrition accomplishing itself in the ferment. It was simply because although Virchow had held that the cellules were living in a living organism, the conception of Bichat was more and more regarded as unacceptable and the hypothesis of the cellularists as unfounded.

Alfred Estor, who was interested in my researches, in giving an account of them in 1865, expressed himself as follows:

> " It is easy to perceive the tendencies of M. Béchamp; each cellule lives like a globule of yeast; each cellule should modify by use the materials of nutrition which surround it, and the general history of the phenomena of nutrition teaches us that these modifications are due to ferments. We know what emotion has welcomed the admirable works of Virchow upon cellular pathology; in the remarkable researches of the Montpellier professor there is to be found nothing less than the foundations of a cellular physiology."[10]

Seven years had passed since the publication of the memoir upon the inversion of cane sugar by moulds, when Estor delivered this judgement and when I wrote to Dumas the letter upon living agents which, in the milk, effect its spontaneous alteration and which, in the chalk, effect the liquefaction and fermentation of fecula starch. The year following I first named the microzymas in the *Comptes Rendus* of the Academy of Sciences to designate the ferments of the chalk.

It has been know since the time of Leuwenhoeck (17th century) that human saliva contains a great number of microscopic organisms long since recognised as vibrioniens, but which in a cleanly kept mouth I have found to be chiefly microzymas.

I supposed that, even as the "little bodies" inverted cane sugar in the experiments of 1857, these microzymas might be those which produced the salivary diastase of Miathe in the saliva. I interested Estor and Camille Saintpiere in this question, and in 1867 we addressed a note to the Academy, having this title: *On the Role of the Microscopic Organisms of the Mouth in Digestion in General, and Particularly in the Formation of the Salivary Diastase.* The note was sent for examination to a commission composed of Louget and Robin, who made no report, and the note was mentioned in the *Compte Rendu* in the following terms:

> "The conclusion of this work is that it is not by an alteration that the parotidian saliva becomes able to digest fecula, but by means of a zymas which the organisms of Leuwenhoeck secrete there, while nourishing themselves upon its materials."[11]

We demonstrated two facts, equally essential; that the buccal microzymas of man liquefy and saccharify the starch of fecula with rare energy; that the parotidian saliva of the dog or horse can also liquefy starch, but does not saccharify it, while such as has stayed upon the buccal organisms soon becomes as saccharifying as human saliva.

The short note inserted by the commissioners shows that they had no idea of a zymas produced as a function of a cellule, of a vibrionien, or of a microzyma, nor even of an organ. Here is an indisputable proof thereof: the pancreas was known and it was called an intestinal salivary gland.

Now Bernard and Berthelot, studying the pancreatic juice and isolating from it the soluble substance called *pancreatin,* never thought for a moment to compare it to the salivary diastase, although it possessed, to the same degree, the power of saccharifying the starch of fecula; that is, Bernard, contrary to the opinion of Longet and of

Mialhe, held that salivary diastase, according to the ideas of Liebig, was an animal matter in a condition of alteration.

The microzymas being discovered, the general demonstration was made that the soluble ferments were substances produced by a living organism, mould, yeast, geological microzyma, diverse flowers, a fruit, the kidneys, and the buccal microzymas. But these were only the preliminary researches, whereof the totality have, since 1867, enabled the microzymian theory of the living organism to be formulated.

After our joint experiment upon the buccal microzymas, I showed Estor an experiment in which a piece of muscle placed in fecula starch, after having liquefied it and commenced to make it ferment, caused bacteria to appear in it as they appeared in soured and clotted milk. He then became my collaborator in proving that that which was true of milk and meat was also true for all the parts of an animal. There has resulted from this, thanks to other collaborations and other researches subsequent to 1870, the microzymian theory of the living organism, the construction whereof is completed by the present work.

The new theory rests upon a collection of fundamental and new facts which may by ranged under the following heads:

1) *The verification of the old hypothesis of atmospheric germs and the ideas of Cagniard de Latour and Schwann regarding the nature of beer yeast.*
   i) Proof that the ferments are not the fruits of spontaneous generation.
   ii) Demonstration that the soluble ferments or zymas are not the products of some change of an albuminoid matter, but the physiological products of a living organism; in short, that the relation of a mould, of beer yeast or of a cellule and of a microzyma with the zymases is that of producer to a product.
2) *The distinguishing of chemical, i.e. not living, organic matters reduced to the condition of definite proximate principles from natural organic matters, such as they exist in animals and plants.*

The proximate principles are naturally unalterable; they do not ferment even when (being creosoted) they are left in contact with a limited quantity of ordinary air, in water at a physiological temperature. On the other hand, *natural organic matters,* under the same conditions or absolutely protected from atmospheric germs, invariably alter and ferment.

3) *Demonstration that natural organic matters are spontaneously alterable, because they necessarily and inherently contain the agents of their spontaneous alteration.*

That is, productions similar to those which I called "little bodies" in certain experiments upon sugared water, and "the living beings already developed," in the letter of 1865 to Dumas, and to which I gave the name of microzymas the following year, as being the smallest of ferments, often so small that they could only be seen under the strongest enlargements of the immersion objectives of Nachet, but which I had discovered to be the most powerful of ferments.

What does this similitude of form and of function mean? What was there in common between a microzyma proceeding from a germ of the air, a microzyma of the chalk, a microzyma of the milk, and those of natural organic matters?

Ever since 1870 all my efforts have been directed to its discovery. My joint researches with Estor, later those of Baltus, upon the source of pus; those of J. Béchamp upon the microzymas of the same animal at its various ages and my own, especially those upon milk, eggs and the blood, have led me to consider the microzymas not only as being living ferments and producers of zymases, like the moulds born in sugared water, but as belonging to a category of unsuspected living beings without analogy, whose origin is the same.

In fact, *first*, all these researches showed me these microzymas functioning like anatomical elements endowed with physiological and chemical activity in all the organs and humours of living organisms in a perfect state of health, preserved there morphologically alike and functionally different, *ab ovo et semine,* in all the tissues and cellules of the diverse anatomical systems, down to the anatomical element which I have called *microzymian molecular granulation.* And especially, they showed me that the cellule is not the simple vital unit that Virchow believed, because the cellule itself has microzymas as anatomical elements.

Secondly, the experiment showed me that in parts subtracted from the living animal, the microzymas, being no longer in their normal conditions of existence, produced therein chemical alterations, called fermentations, which inevitably led to tissue disorganizations, to the destruction of the cellules and to the setting free of their microzymas, which then, changing in form and function, could become vibrioniens by evolution, which they did whenever the conditions for this evolution were realized.

And, *thirdly,* I established that the vibrios, the bacteria which the anatomical microzymian elements had become, destroyed themselves, and that, with the aid of the oxygen of the air, under the conditions which I had realized, they were at last reduced to microzymas while the matters of the alteration, being oxidised, were transformed into water, carbonic acid, nitrogen, etc, i.e. they were restored to the mineral condition, so that of the natural organic matters and of their tissues and cellules there remained only *the microzymas.*

These microzymas, proceeding from the bacteria which the anatomical element microzymas had become, were identical, morphologically and functionally, with those of chalk, calcareous rocks, alluviums, water, arable or cultivated earths, or the dusts of the streets and the air. From these experiments, I argued that the microzymas of the chalk, etc, were the microzymas of the bacteria which the anatomical element microzymas of the living beings of the geological epochs had become!

We then have to consider:

1) The microzymas in their function as anatomical elements in the living and healthy organism; there they are the physiological and chemical agents of the transformations which take place during the process of nutrition.

2) Microzymas in natural organic matter abstracted from the living animal, or in the cadaver; there they are the agents of the changes which are ascertained to take place there, whether or not they undergo the vibrionien evolution—changes which lead to the destruction of the tissues and the cellules.

AUTHOR'S PREFACE / 1

3) The microzymas of the bacteria which result from this evolution, which are essentially ferments productive of lactic acid, acetic acid, alcohol, etc, with sugar and fecula starch; these microzymas are also producers of zymases and are capable of again undergoing vibrionien evolution.

The microzymas being the anatomical elements of the organized being from its first lineaments in the ovule which will become the egg, I am able to assert that *the microzyma is at the commencement of all organization*. And the microzymas of the destroyed bacteria being also living, it follows that these microzymas are *the living end of all organization*. The microzymas are surely then living beings of a special category without analogue.

But that is not all. Estor and I demonstrated that in a condition of disease, the microzymas which have become morbid determine in the organism special changes, dependent upon the nature of the anatomical system, which lead alike to the disorganization of the tissues, to the destruction of the cellules and to their vibrionien evolution during life, so that the microzymas, living agents of all organization, are also the agents of disease and death under the influences which nosologists specify.

Finally, they are the agents of total destruction when the oxygen of the air intervenes. Like the indestructible atom or element in the Lavoisierian theory of matter, the microzymas, too, are physiologically imperishable.

From the experimental fact that the microzymas of the chalk and dusts of the air are only microzymas from bacteria which proceeded from the vibrionien evolution of the anatomical element microzymas, it follows that that which I have called *germs* in my verification of the old hypothesis of *germs of the air* are *not* pre-existent in the air, in the earth and in the waters, but are the *living remains* of organisms which have disappeared and been destroyed.[12]

The facts of the microzymian theory have legitimatized the genial conception of Bichat; that the only thing living in an organism is what he regarded as elementary tissues. Later, among cellularists, Virchow, following Gaudichaut, held that the cellule was the simple anatomical element from which proceeded the whole of a living being; but it is in vain that he contended that it is the vital unit, living *per se*, because every cellule, even that of beer yeast, *is transitory*, destroying itself spontaneously.

It is the microzyma which enables us to specify precisely wherein a tissue, a cellule is living; living *per se*—that is to say, autonomically, it is in truth the simple vital unit.

But the conception had none the less as a consequence the assertion that, in disease, it is the elementary tissues or the cellules which are affected.

Tissue and cellular physiology now being established in accordance with the prevision of Estor, it should result from this that tissue and cellular pathology are in reality microzymian pathology.

In disease, the cellules have been seen to change, to be altered and destroyed, and these facts have been noted. But if the cellule were the vital unit living *per se*, it would know neither destruction nor death, but only change. If then the cellule can be destroyed and die, while the microzyma can only change, it is because the microzyma is really living *per se*, and physiologically imperishable even in its own evolutions, for, physiologically, *nothing is the prey of death;* on the contrary, experience daily proves that everything is the prey of life, that is to say, of what can be nourished and can consume.

From the beginning of our researches, Estor and I have established the presence of microzymas in the vaccine matter, in syphilitic pus as in ordinary pus, and I have shown in pus (even laudable) the presence of a zymas. In diseases there is, then, a morbid evolution of some anatomical element which corresponds to a vicious functioning and to vibrionien evolution.

It is thus that in anthrax the morbid microzymas of the blood become the bacteria of Davaine, and the blood globules experience such remarkable changes. But even as the microzymas may become morbid, they may cease to be so. For instance, there is a leading observation of Davaine upon the non-transmissibility of anthrax even by inoculation; if the animal is in process of putrefaction, its blood can no longer communicate anthrax.

From this observation of Davaine, I draw the conclusion that normal air never contains morbid microzymas, or what used to be called germs of diseases and are now called microbes; maintaining, in

AUTHOR'S PREFACE / 1

accord with the old medical aphorism that *diseases are born of us and in us,* that no one has ever been able to communicate a characteristic disease of the nosological class (anthrax, smallpox, typhoid fever, cholera, plague, tuberculosis, hydrophobia, syphilis, etc.) by taking the germ in the air, but necessarily from a patient, at some particular moment. And within the limit of my own studies upon the silkworms I distinguished with care the parasitic diseases whereof the agent came from outside, such as the muscardine and the pebine, from constitutional diseases, such as the flacherie, which is microzymian.

I give in the postscript of this work the communication which I made to the Academy of Medicine on the 3rd May, 1870, upon *Les Microzymas, la Pathologie et la Therapeutique.* It will help to establish the date, and will show that the theory was then nearly complete. It was not inserted in the Bulletin of the Academy, but an able physician, who gave an account of it in the *Union Médicale* of Paris, remarked that had it come from Germany it would have been received with acclamation. But there was not at that time any question about the medical doctrines of Pasteur and I did not then have to defend the microzymas against the denials of that savant; it was otherwise some years later.

The foregoing exposition shows clearly the connection of the new facts of the microzymian theory with certain earlier facts of the same kind, ascending to Bichat and Macquer, who, in agreement with the science anterior to Lavoisier, recognized the spontaneous alterability of natural organic matters; and at length Spallanzani, who, to explain certain apparitions of organized beings ascribed to spontaneous generation, invoked the germs of the air. It has enabled me further to follow the connection of the successive discoveries of special facts which, since 1854, the commencement of these researches, have resulted in the discovery of the microzymas and to the demonstration that the blood is a flowing tissue.

It is important to remark that the microzymian theory is in no way the product of a system or of a conception *a priori,* nor is it the consequence of a desire to demonstrate that the conception of Bichat and the cellular theory are conformable to nature. In fact, it has had for a point of departure the solution of a problem of pure chemistry and the necessity of discovering the role of the moulds in the inversion of a solution of cane sugar exposed to the air. Then, from induction to induction, applying unceasingly the method of Lavoisier, and from the attentive study of the properties of the lowest organism, I ascended to the highest summits of physiological chemistry and of pathology to discover wherein vital organization consists.

But so fertile is this theory founded upon the nature of things, and which has as its base no gratuitous hypothesis, that after it had led me to discover the source of the zymases, the physiological theory of fermentations, the nature of what were called the germs of the air, it enabled me to understand what was true in the ideas of Bichat, Dumas, and in the cellular pathology of Virchow and what profound truths there are in the aphorisms of the old physicians.

The microzymian theory of the living organism is true because it agrees at the same time with these conceptions and with the three aphorisms which I have chosen as the epigraph to this first part of my preface.

Nothing is but what ought to be.
Nothing is created; nothing is lost.
Nothing is the prey of death;
all things are the prey of life.

NOTES

1. L. Pasteur, *Annales de Chimie et de Physique*, 3rd S. vol LVIII, p.371, note.
2. It is wrong to suppose that the word organic, in *organic* molecules, had the same meaning as in *organic matters* of modern chemists; this is so little the truth that Buffon admitted *organic* molecules to explain the crystallisation of marine salt or of others, purely mineral.
3. A study already made by Desmazieres, who regarded the globule of beer yeast as an infusoria under the name of *Mycoderma cerevisiae,* but which Turpin called a plant under the generic name of *Torula* and Kutzing under that of *Cryptococcus.*
4. [The translator bespeaks a painstaking attention by men of science, by philosophers and by philanthropists to the rest of this narrative; and to keep in mind the constant boastings by literature, by the press, and by men held as most eminent in science of our superiority over our fathers. Can that superiority be proved to exist elsewhere than in the arts of murder, and what pertains thereto?—*Trans.*]
5. [Here is the discovery and source of all that is true in the theory and practice of antisepsis; but it has been carried to absurd extremes by the dominant faction in medicine.—*Trans.*]
6. *Annales de Chimie et de Physique*, 3rd S., Vol. LIV, p.18 (1858).
7. I have called by the name of moulds the totality of the productions which have appeared in different experiments which I have diversified. Generally these moulds remained in the state of colourless mycelium, even in solutions to which arsenious acid and certain salts had been added. In others, the completely developed mould was green or gray and rarely red. In some experiments there were actual cellules, different both from yeast globules and from the ferment of the lees of wine. Generally, at the beginning of the experiment, there was a slight deposit before the appearance of the mycelian tubes; in some cases, the inversions were effected by the "isolated little bodies," *little bodies* which I had not known how to classify, but which I held to be organized and living because, like the mycelian moulds, they effected the inversion of the sugar even in a creosoted solution.
8. Letter to Dumas, *Annales de Chimie et de Physique*, 3rd S., Vol. VI, p.148 (1865).
9. *C.R*, vol. LVIII, p.601 (4 April, 1864).
10. Montpellier, *Le messager du Midi (*1865).
11. *C.R.*, Vol. LXIV, p.696 (1867).
12. [The "experimental fact" referred to in the text (the very highest form of all evidence which can be supplied by science) cuts away the entire fabric of the microbian theory of disease from its very foundation. Never having been other than a baseless *guess* on the part of Pasteur and of his followers, it was fittingly designated by Béchamp as "the greatest scientific silliness of the age." It and the other "experimental facts" learnedly elaborated by Professor Béchamp and his collaborators make patent the absurdity of all pretended prophylactics against disease save one, and casts all rational minds back to the one sure and only

protection—sound hygiene!

We are mocked by quarantines, vaccines, inoculations and other devices for "conveying" the products of labour into the pockets of official doctors. We are gulled by them to the full extent of our willingness to be gulled. The opponents of a truly rational medicine are many and powerful, as evidenced by the suppression for more than a generation of Béchamp's admirable discoveries beneath a conspiracy of silence, and these opponents of the art of healing are entrenched in nearly all medical schools, in richly endowed research institutes, in expensive manufactories of animal poisons for poisoning men and animals (under the ignorant belief that they are benefiting us), and in all medical officialdom! -*Trans.*]

AUTHOR'S PREFACE / 2

> "The greatest disorder of the intellect is to believe things because one wishes that they were so."
>
> – *L. Pasteur*[1]

To understand how man's intelligence, arrested at the same stage that it was in the days of Aristarchus, could come to proscribe the microzymian theory of the living organization as it had proscribed the theory of the movement of the earth, it is necessary to know something of the prejudices with which man's intelligence in these latter days has been imbued.

The Lavoisierian theory of matter suggested to Bichat the idea that in organized beings, life is not connected merely with chemical compounds, but also with anatomical elements personally and autonomically living. This caused Fourcroy to say that plants are organized machines which formed the matters extracted from them, matters which Chevreul will call *definite proximate principles,* and which no instrument of art is able to imitate. Gerhardt in 1849 will say of them that they are the work of the *vital force*. It was in vain that Bethelot, therein recalling Lavoisier, will think to prove that the proximate principles are chemical compounds such as those whose synthesis he effected; all the legitimate consequences of the conception of Bichat were disregarded, even the notion that the cellule is personally living, and it was maintained that:

> " The proximate principles of plants and animals are bodies, definite or not, generally very complex, gaseous, liquid, or solid, constituting organized substance by reciprocal solution, viz.: the humours, and by special combination, the anatomical elements."[2]

'Reciprocal solution' and 'special combination'; vague expressions used to conceal a preconceived system, thanks to which it was only necessary to consider the proximate principles in a living organism as purely chemical matter.

The autonomous nature of the anatomical elements in the tissues being thus set aside, it was declared that the protoplasm of the botanist Hugo Mohl was *living, organized matter* (although not morphologically determined, that is to say, not structured), whence the entire organism would proceed. It was thus that a liquid, in which all the proximate principles were supposed to be in a state of perfect solution, such as was called *plasma* in the blood, was called organized, living, and could die.

This was going back beyond the hypothesis of organic molecules of Buffon to the old hypothesis of matter living by its nature and to that of an organization which would be only the most excellent modification of matter such as it was imagined to be in the epoch of phlogiston.

That is where science stood in 1857; seeing in animal membranes and tissues only nitrogenised matter. Let us consider the consequences of this mode of view.

In 1839, Fremy found that certain animal membranes could produce lactic acid with the sugar of milk, which Scheele had discovered in the whey of soured and clotted milk. Thereupon lactic fermentations were produced by treating solutions of the sugars with all sorts of animal membranes and tissues, with cream cheese or gluten, and at the same time with chalk used to saturate the lactic acid as it was produced.

Berthelot resumed these experiments from another point of view, without neglecting the formation of lactic acid, but extending it from mannite sugar to allied substances, even to glycerine. The memoir wherein, in 1857, the author explained the results of his researches is entitled *Sur la Fermentation Alcoholique,*[3] for it happened that in some cases the quantity of alcohol formed was greater than that of the lactic acid and other products which accompany them.

But whatever name may be given to the phenomenon, lactic or alcoholic fermentation, that which resulted from the experiments of Berthelot was that:

> " The cause of fermentation seems to reside in its chemical nature; that is to say, in the composition and not in the form of the nitrogenous bodies (cream

cheese, yolk of egg, muscle, pancreas, liver, kidney, spleen, testicle, bladder, small and large intestines, lung, brain, hairy skin, blood, dried fibrin, dried yeast, gluten, gelatine) fit to play the part of a ferment, and in the successive changes which their composition undergoes."

On the whole, he was of opinion that:

"The sugared body and the nitrogenised body are decomposed at the same time, exerting upon one another a reciprocal influence."

In short, it was spontaneous fermentation of materials in the presence of one another.

As to the chalk employed for calcic carbonate, it was supposed to be absolutely needed only in certain cases, for example for the fermentation of mannite; further, the calcic carbonate, besides maintaining the neutrality of the medium, had for its role:

"to direct in a certain determined sense the decomposition of the nitrogenised body which provokes the fermentation."

So far as an explanation of the phenomenon went, Berthelot seemed to relate it to the saccharification of fecula by diastase, the decomposition of amygdalin by synaptase, called fermentation, or even the etherification of alcohol by sulphuric acid; in short, to connect it, as did Mitscherlich and Berzelius, with an action called *catalytic contact*.

Berthelot did not fail to have established by Robin, Montagne, and Dujardin, the disorganization of the tissues and the development of particular living beings (mucors and vibrios or bacteria). He does not explain their source, makes no mention of the molecular granulations, but, he asserts, "this development is in no way necessary to the success of my experiments."

I have endeavoured to give an idea of the very important work of Berthelot because it constitutes the greatest effort in opposition to the opinion of Cagniard de Latour. But from the same experiments, entirely contrary conclusions ought to be drawn.

In fact, the following year Pasteur, in a memoir upon lactic fermentation[4] of sugar, under the conditions of Berthelot's experiment, placed himself on the side of Schwann and asserted that the development of special living beings was the sole cause of the fermentations pointed out, but without paying any more attention to the molecular granulations that Berthelot had done, he had the merit to distinguish among the particular living beings that which he named *lactic yeast,* and which he regarded as being to lactic fermentation what beer yeast is to the alcoholic.

But of the development of these beings, especially of the lactic and alcoholic yeasts, what according to him, was the cause? He had the choice between two hypotheses; that of the germs of the air with Spallanzani and Schwann, and that of spontaneous generation; he chose the second, asserting that these beings were born spontaneously of the albuminoid matter of the nitrogenised matters. To prove this he made the two following experiments which are important to remember:

> " The lactic yeast is born spontaneously with as much facility as beer yeast wherever the conditions are favourable.
>
> Let there be, for example, first, water of sweetened yeast without addition, and, second, the same with the addition of chalk.
>
> In the clear solution of the first we have beer yeast and the alcoholic fermentation; in the solution to which chalk has been added it is lactic yeast and lactic fermentation which will be developed. The yeasts are born spontaneously of the albuminoid matter furnished by the soluble part of the yeast; the beer yeast because the water of the yeast is acid, the lactic yeast because the chalk makes the yeast neutral."

We can say then that Pasteur and Berthelot have proposed, each in his own way, the spontaneous alteration of nitrogenised matter under the conditions specified by Macquer, but while this alteration resulted in the spontaneous generation of the ferments according to Pasteur, Berthelot did not express his views upon the origin of the living beings developed.

As to the manner in which the lactic yeast acted, how did Pasteur understand it? Cagniard de Latour had said that the fermentation of the sugar was an effect of the vegetation of the yeast; Pasteur said of the lactic yeast that "its chemical action is correlative of its development and of its organization", which, though in other words, is the same thing and may be classed as an explanation by catalytic contact.

I have insisted thus strongly upon this earlier work of Pasteur upon fermentations for two reasons:

First, to firmly establish how vain had been the efforts of Schwann to establish the idea that there can be no spontaneous alteration of organic matters by fermentation without the presence of special living beings, and that in conformity with the hypothesis of the germs, these living beings were not the product of spontaneous generation.

Secondly, to show how in 1858 Pasteur, having remained a sponteparist with regard to these living beings and as to beer yeast and lactic yeast, held that these organic matters were spontaneously alterable. We shall see how some years later Pasteur will 'discover' all of a sudden that ferments are never born spontaneously, but always from these atmospheric germs which he had neglected; he will even 'discover' that albuminoid matter is not necessary for it. He will next pretend to demonstrate that without these germs all organic matter, without exception, even an entire cadaver, will remain unchanged indefinitely. First it will be useful to know certain parts and certain conclusions of his memoir upon the alcoholic fermentation of cane sugar by beer yeast in the year 1860.[5]

From this work it is first to be remembered that Pasteur in it again asserts the spontaneous generation of beer yeast and then the fact, absolutely new, that glycerine is among the products of fermentation, the same as in wine of vinous fermentation. He also discovered in it succinic acid, which had been long before discovered in it by Schmidt.

With regard to the chemical action of the cellule of beer yeast, it is equally correlative with its development and organization. He was, in fact, so certain that the yeast took no other part in the phenomenon that he laboured hard to prove that all the products of fermentation came from the sugar, which would be a physiological heresy if fermentation is a phenomenon of nutrition which is accomplished within the ferment.

It is thus that upon the interesting question of whether the cane sugar ferments directly, or if it is first inverted (as was the opinion of Dubrunfaut in agreement with the remark of Dumas, who had shown that for the equation of fermentation the concurrence of water with the cane sugar is necessary), Pasteur pronounced for direct fermentation, asserting that the inversion was consecutive to the formation of succinic acid.

Nevertheless he knew that I had demonstrated the inversion of the sugar by organized productions which are born in sugared water exposed to the air. None the less he wrote the following, which is typical: "I do not think that there is any special power in the globules of yeast to transform cane sugar into grape sugar."[6]

He knew also that Berthelot had supposed that the reduction of the sugar into alcohol and carbonic acid was to be compared to the reduction of amygdaline by synaptase. He knew that Dumas had clearly stated that yeast, like an animal, could not be nourished only upon sugar; that for its normal life an appropriate albuminoid matter was needed. If he did nothing to elucidate these important questions it was because he was obsessed with the preconception that there is nothing in common between the organization and life of a cellule of yeast and that of an animal cellule. This was because he regarded it as certain that the ferments are living beings apart by destination, and that fermentations are individual phenomena. He asserted that a special ferment corresponds to each fermentation.

This state of mind and a remark suggested to Pasteur an experiment which Doctor E. Roux, wonderstruck, called an *"experiment a la Pasteur."*

This memorable experiment had for its object the multiplication, that is to say, the vegetation with reproduction, of beer yeast in a sugared medium without the addition of some appropriate albuminoid matter. The remark which made him attempt it was as follows:

Pasteur had been greatly impressed by the results of my experiments regarding the inversion of cane sugar by the various productions which are developed in its aqueous solution, and especially by the fact that the addition of certain non-ammoniacal mineral salts had the effect of increasing the harvest of these productions while causing them to vary. Now the nitrogen necessary for the synthesis of the albuminoid matters of these moulds could only have been that of the air left in the flasks in contact with these sweetened liquors.

Pasteur repeated the experiments and was convinced not only that true ferments of many species were developed without the employment of albuminoid matters, but that these ferments had formed these matters by synthesis. Then he who had asserted that the ferments were spontaneously born from the albuminoid matters of the sugared media had to amend his former opinion.

Assuredly, no more than I, could Pasteur have seen the beer yeast appear under the conditions in which the experiments had been reduced to their simplest expression, in order to make more strikingly plain the evidence that there could be no question there of spontaneous generation.

He thought he would succeed better by adding to a solution of candied sugar the right tartrate of ammonia and for mineral salts the ashes of the yeast itself; he succeeded no better, then he added to the same mixture a lot of yeast, in the hope that the tartrate of ammonia and the sugar would form by copulation an albuminoid matter which would help the multiplication of the globules of yeast. There are two versions of the results of the experiments.

One, that of Roux, more or less agreeing with or imitated from an earlier one of Pasteur, is the following: "Pasteur," he said, "had seen carbonic acid set free, the yeast augmented ... he observed that *all the sugar had disappeared,* transferred into alcohol, carbonic acid, etc."[7]

The other, by Pasteur,[8] is very different from that. There was set free, in fact, carbonic acid, but in microscopic globules; some sugar had disappeared, but out of ten grams, 5.5 grams had not fermented: there was some alcohol, but only a very small quantity, sensible but not sufficient to weigh, etc. What then had become of the sugar that had disappeared? It had become lactic acid, which had furnished "an abundant crystallisation of lactate of lime"; in short, the fermentation instead of being alcoholic had been lactic!

Now for the explanation of the facts according to the microzymian theory:

Pasteur, having continued to neglect the hypothesis of germs, found that the situation of the beer yeast being extra-physiological, its globules had proliferated at the expense of the reserve of their content, so that the time soon arrived when these were *exhausted,* the new after the old, while infusoria and lactic yeast overspread the liquor. "The infusoria disappeared and the lactic yeast multiplied," said Pasteur. About a month later, the lactic yeast 'continuing to increase', the ferments were collected and weighed.

Pasteur gave his results as being "of the most rigorous exactness." I, however, assert that under the conditions of his experiment, the quantity of yeast collected *must* have been less than that of the yeast sown. Now, reflecting upon what he thought was an increase of the yeast and this production of lactic yeast, he has given this experiment "as illuminating with a new day the phenomena of fermentation."

This declaration is applicable to my experiments of the memoir of 1857, which are really demonstrative and which Pasteur has attempted to ascribe to himself while imitating after repeating them. In fact it was a plagiarism to the detriment of science.[9]

To complete the exposition of the state of the question in 1860, here is an experiment of Berthelot in the sense of mine. The author made a solution of gelatine, of glucose and of bicarbonate of potash, saturated it with carbonic acid, filtered it while warm in a still which he filled completely and left to itself. At the end of a greater or less time (some weeks) gas was set free and a good deal of alcohol was formed. At the same time a slight, insoluble deposit was formed "composed of an enormous number of molecular granulations, much smaller than beer yeast and very different in appearance." [10]

Berthelot did not ascribe any role to these molecular granulations, and believing that he had performed the experiment "protected from contact with air", he asserted, as in 1857, that the presence of calcic carbonate (the chalk) or of any alkaline bicarbonate directs the decomposition of the nitrogenised body (in this instance, the gelatine) in a certain manner which sets up the fermentation by regulating the steps of the phenomena. In short, Berthelot had not yet distinguished between the calcareous rocks (the chalk) and pure calcic carbonate, exactly like Pasteur in this matter, and did not yet believe that atmospheric germs had anything to do with the appearance of the molecular granulations. In short, he naturally believed that the lactic yeast of Pasteur was also constituted of molecular granulations, and that there was nothing to show that it was organized and living; as was the opinion of Pasteur, who, in 1858, stated that he had argued "*on the hypothesis* that the new yeast was organized and living."

This, then, was the state of knowledge in 1860, and even much later. It was not known, although it already stood out from the facts of my memoir of 1857, and which the microzymian theory has since confirmed, that that which characterizes the fact of a living organization is not essentially, as the naturalists of the schools still believe, the establishment of the existence of some organ or structure, nor is it the presence of movement more or less spontaneous or voluntary in any living being whatever, or such as a microzyma, molecular granulation or lactic yeast, or such as a vibrionien. Rather, living organization is characterized by the property of producing and secreting zymases, each according to its nature or species; and the production of the chemico-physiological phenomena of transformation called fermentation, which are acts of nutrition, that is to say, of digestion, followed by absorption, assimilation, disassimilation, and so forth, and finally, the ability to reproduce itself if all conditions dependent upon nutrition are fulfilled.

This is what Pasteur could not understand when he alleged in 1860 that the fermentation of cane sugar by beer yeast was correlative to the multiplication of the yeast, which is as great a physiological heresy as to imagine that an animal could be nourished upon sugar alone.

But soon after, Pasteur, who had not yet explicitly invoked the germs in explanation of the alterations of organic matters and the production of the alterations of organic matters and the production of ferments, would explain by them what he had before explained by spontaneous generation; in short, he held my verification of the hypothesis to be so rigorously correct that in 1862 he published a memoir against spontaneous generation, wherein the alteration of all organic matters was explained as Schwann had done, by applying his method as improved by Claude Bernard.

That was his second plagiarism.

His experiments in the memoir of 1852 had been made with the organic substances treated, cooked, for the purpose of killing the germs which the air might have deposited upon them. In 1863 he repeated them upon blood and flesh, not cooked, for the purpose of proving that they did not contain germs capable of becoming vibrios, and that, without atmospheric germs, they would be unalterable. Not being able to heat flesh in the same manner as blood, he applied my method, substituting alcohol in the place of creosote. That was a third plagiarism. But he could not see the vibrioniens which, in spite of the antiseptic agent, were developed in the depths of the flesh, and he concluded that neither the blood nor muscle became putrid because the germs of the air were absent from them. And he regarded as proven that there was nothing living in the blood or in the flesh, and that all animal matters, without the germs of the air, would remain indefinitely unchanged.

While Pasteur thus experimented, I continued to develop the consequences of my memoir of 1857. I demonstrated especially that not only were the atmospheric germs unnecessary for vinous fermentation, but that they were injurious, and that the grape carried normally, upon itself, the cellules of the ferments of the lees; not only the germs but the fully developed ferment. This was in 1864.

At last, in 1865, I announced to Dumas the fact of the existence in the milk, and in the chalk, of the agent which is the cause of the spontaneous alteration of the former and of that which enables the second to act as lactic ferment, agents to which in the following year I gave the name of microzymas.

Pasteur, who had been named a member of the commission[11] upon my memoir upon the ferment of the chalk, said not a word, and I continued with Estor the study of the microzymas of the higher organisms up to applications to pathology, as may be seen in the postface. This was in 1870.

In 1872 Pasteur attempted his boldest plagiarism; he discovered all of a sudden, eight years after my discovery thereof (I will state elsewhere on what occasion), that the ferment of vinous fermentation exists naturally upon the grape. In this connection he discovered also that plant and animal matters contain normally the things which cause them to alter spontaneously; that their cellules, without the atmospheric germs, are ferments. In other words, he repudiated his experiments and conclusions of 1862. He announced that his *'new discoveries'* would mark an epoch in general physiology; and he asserted that he had thrown a great light upon the phenomena of fermentation and had "opened a new path to physiology and medical pathology."

This was too much: up till that time I had treated the man with consideration; but now he must be properly exposed. First I, then Estor and I together, protested energetically. Our protests were inserted literally by Dumas and by Elie de Beaumont; the complete text can be read in the *Comptes Rendus*, Vol. LXXV, pp1284, 1519, 1523 and 1831. Pasteur replied by a subterfuge, to which we replied as follows: "We request the Academy to permit us to record that the observations inserted in the names of M. Béchamp and of ourselves remain unanswered".

Pasteur said no more, and abandoning "the new road" he pretended to have opened (a road which we showed we had not only opened but had sturdily traversed) he retraced his steps. Then, while since 1858 he had not disputed the meaning of any of the results, of any of the facts upon which the microzymian theory rests, results and facts which he knew to be exact and the discovery whereof he tried to ascribe to himself; then, I say, it was that he undertook in 1876 to explain them all by the atmospheric germs as he had "explained" them, in 1862, by spontaneous generation.

He first evoked his experiment upon the blood in 1863, and, doubtless because Estor and I, after the discovery of the microzymas of the fibrin, had not thought it worth criticizing, he qualified it as *famous(!)*, using it to deny even the existence of the microzymas. He then canvassed for approvers to maintain that uncooked milk, like the blood, is unalterable when preserved from contact with the natural air; that without atmospheric germs there would be neither fermentation nor disease, because there would be neither ferments nor *microbes*; for Pasteur, in spite of the inaccuracy of the etymology, had adopted this word with which to designate the micro-organisms.

In short, Pasteur, who understood what he was about in this matter, ended by causing belief that things were as he wished they were, which as he himself had said, "is the greatest derangement of the mind."

The strangest part of the business is that it was believed, and that he was able to make the Academies his accomplices.[12] It is true that he had at the same time organized the conspiracy of silence around the works related to the microzymian theory—so thoroughly, that one day, after a discussion during which I had attacked the principles of the microbian doctrines and had defended the microzymian theory, Cornil maintained that the discoveries of Pasteur had been verified in every country and that I was alone against all the would; to which I replied:

> " It is not because everybody thinks so that it is true. I have demonstrated in an already old communication that the protoplasmic system, false in its principles, is false also in its consequences. It is so likewise with the microbian doctrines. For the dignity of science and of human reason it is time that they were abandoned!" [13]

The discussion did not rest there. I will narrate the rest, which is most instructive, in *The History of the Microbian Doctrines*, to show the sort of respect which Pasteur had for truth.

It is true we have not been treated as was Galileo by the Inquisition, but Estor, painfully afflicted, wrote me this, which constitutes a grave witness against the spirit of these times:

> " We can publish letters from members of the Institute begging us in the name of our personal interest to proceed no further in the road opened (by us) ... but let them be convinced that energetic protests[14] will be directed wherever one may hope to find associated science and honesty."

That honourable and conscientious savant died of grief!

The microzymian theory has experienced in our days, as was the case formerly, the fate of all new truths which go counter to the habits, the passions, and the interests of those in power.

It is because man's reason, that is to say, that part of it which has become vacillating, without ballast, hypocritical and pharasaical, has remained the same as it was in the days of Aristarchus, of Socrates, of Galileo. It is that part of mankind which allows the plagiarist to calumniate and to vilify the victim whose work he has plagiarised.

NOTES

1. *C.R.*, Vol. LXXX, p.91 (1875).
2. *Dictionnaire de Médecine,* Littré and Robin, articles *Immédiat* and *Organique*.
3. *Annales de Chimie et de Physique,* 3rd S., Vol. 2, p.322.
4. *Annales de Chimie et de Physique,* 3rd S., Vol. LII, p.404.
5. *Annales de Chimie et de Physique,* 3rd S., Vol. LVIII, p.323.
6. *Loc. cit.*, p.357. In relation to this, an observation is necessary. Some persons, not well informed, ascribed to Berthelot the discovery of the property of beer yeast to invert cane sugar. That savant had nothing to do with it. The following is the truth. In 1840, Mitscherlich discovered that the clear liquor obtained by leaving beer yeast to drain upon a filter possesses the property of converting cane sugar into uncrystallisable sugar, whereas the globules of the ferment, well washed with pure water, *are entirely deprived of this property*. And Berzelius added: "*The formation of the uncrystallisable sugar is not due to the globules of the ferment, but to a soluble matter in the water with which they are mixed.*" Now, in 1860, Berthelot had simply confirmed the fact and isolated the soluble matter, whereof Berzelius spoke, but had not demonstrated that there was a special property of transformation of the cane sugar in the globules. That is what I demonstrated after having discovered that the moulds born in the sugared water without albuminoid matter possess individually the inverting power, and it was that which was needed to prove that the soluble ferment was not a product of the change. See *Les Microzymas,* pp.45-47, and *Memoir sur les Matieres Albuminoides,* p.352, for the complete history of the zythozymas.
7. *Revue Rose,* Vol. X, 4th S., p.834 (1898).
8. *Annales de Chimie et de Physique,* 3rd S., Vol. LVIII, pp.383-392 (1860).
9. (From p.42) Roux, evidently for the purpose of causing people to believe in the priority of Pasteur in this matter, has stated that the experiment was made in 1856, anterior to the publication of my memoir, while it really was of the 10th December, 1858, several months subsequent to the publication of the memoir, wherein Pasteur asserted that the ferments were the results of spontaneous generation from albuminoid matters (*prennent spontanément naissance des matieres albuminodies*), posterior by a year to the deposit of my memoir with the Academy of Sciences, published by extracts in the first *Compte Rendu* of 1858, and *in extenso* in the *Annales de Chimie et de Physique* in September of the same year. It was in the same spirit, that before that time, Roux had the audacity to write that "the medical work of Pasteur began with the study of fermentations" (*Agenda du chemiste* for 1896); this was an absolute untruth, for seven years later Pasteur had not yet attained to an elementary understanding about them; Roux either did not go to the original documents or he was anxious to contribute to the legend which attributes to Pasteur the discovery of the facts of the microzymian theory. That legend is a falsehood.

[A further "illumination" is thrown upon this subject, so discreditable to science and its professed masters during the last quarter of the nineteenth century, in *Les Grands Problemes Médicaus,* Paris, 1905, pp12-13. The statements wherein can be verified by anyone possessed of a moderate knowledge of physiological chemistry, who will take the trouble to read, and to study, Section III of Pasteur's memoir to be found in the *Annalse de Chimie et de Physique,* 3rd Series, Vol. LVIII, pp381. On such study being made, the experiment there pretended to have been made will be seen to be A FAKE, purely and simply.—*Trans.*]

10. *Chimie Organique Fondée sur la Synthese*

11. [When a memoir is presented to the Academy which seems to be of more than usual importance, a commission is named composed of members reputed skilled in like studies to examine and report upon the memoir. It was of such a commission on the memoir of Béchamp upon the chalk ferments whereof Pasteur was appointed a member.—*Trans.*]

12. The following is typical in this respect. Pasteur had treated Fremy shamefully, because he had maintained that cream cheese produced lactic fermentation of itself. I said to him: "But show then to the Academy the microzymas of the milk and of the cream, which are the lactic ferment of Pasteur and you will confound him." "Ah," said he, "I should never dare to pronounce the word *microzyma* at the Academy." To such an extent indeed had Pasteur cunningly manoeuvered!

13. *Bulletin de l'Académie de Médecine,* 2nd S., Vol. XV, p.379 (1886).

14. [The translation of this work, and its publication, is one of the first of those *protests* which Estor foretold. It is hoped that it will mark the turning point of the followers of science from the wisdom of the "philosophers of Lilliput," in which so many of them have been wallowing—and, what is worse, training students of biology, physiology, pathology and medicine to mistake follies for wisdom! -*Trans.*]

AVANT PROPOS

The object of this work is the solution of a problem of the first order; to show the real nature of the blood, and to demonstrate the character of its organization. It has, besides, a secondary purpose; the solution of a problem long ago stated, but never solved—the cause of its coagulation, correctly regarded as spontaneous, after it has issued from the blood vessels.

The conclusion arrived at is that the blood is a flowing tissue, spontaneously alterable in the same manner as are all other tissues withdrawn from the animal, coagulation of the blood being only the first phase of its spontaneous change.

It would be too tedious to give even a summary of what had been written upon the blood before the discovery by Harvey and that of the blood globules; I will merely observe here that both before as well as after these memorable discoveries, the blood has been almost exclusively called a liquid by those physiologists who specially studied it. This will appear abundantly from the historical introduction, especially with regard to the attempts at explanation of the phenomena known as spontaneous coagulation.

Every year since 1860 at the University of Montpellier, at the commencement of the course on medical chemistry of the Faculty of Medicine, the assistant wrote on the bulletin board an announcement of the fundamental principles of the instruction which would be given by Professor A. Béchamp.

This announcement is included here to demonstrate that already, in 1860, Béchamp's views on the subjects mentioned were settled, and nothing has since occurred to show them to be erroneous:

There is only one chemistry. Matter is endowed only with chemical and physical activity.

There is no matter essentially organic, all matter is mineral.

That which is called organic matter is only mineral matter, with carbon as a necessary constituent.

Organic matter, chemically definite, is profoundly distinct from organized matter.

The chemist can, by synthesis, form organic matter, but he is powerless to organize it; he cannot create a single cellule.

The faculty of organizing matter resides, primordially, in pre-existing living organisms.

It is in the various mechanisms of the organism of organized beings wherein are accomplished the changes of organic matter, whether organized or not; and these changes are effected according to the ordinary laws of chemistry.

From the chemical point of view, plants are essentially apparatus of synthesis, animals apparatus of analysis.

INTRODUCTORY AND HISTORICAL

The explanation of the fact of the coagulation of the blood, rightly regarded as spontaneous, has been sought for by physiologists, by physicians and by chemists, but without satisfactory result. The detailed history of the attempts at explanation would only demonstrate the uselessness of the preconceived hypotheses and systems on which they rest.

Among all these hypotheses, only one deserves attention; precisely the one which the latest investigators have neglected to consider or to verify. The history of the conception of this hypothesis is of great interest.

From time immemorial, it has been known that shed blood soon becomes a concrete mass, red, of a consistency more or less soft and called a clot; the phenomenon was otherwise compared to the coagulation of a homogeneous liquid.

It was not until the 18th century that Haller (in the supplement to the article on blood in the *Encyclopaedia of Diderot*), after correcting some errors of Leuwenhoeck concerning blood globules, asserted absolutely that they were essential elements of the blood, existing only in the red part, and, said he, "perhaps also in milk."

However, he recognised that

> " the figure of the blood globules is constant and that they are not merely a collection of fatty grains ... but are circumscribed, bounded and solid."

Haller also first placed the spontaneous coagulation of the blood on its true ground, tracing the theory back to Aristotle:

> "An element of the blood, generally so regarded by the ancients, especially by Aristotle, are the fibres which the scholiasts regarded as the foundation of the coagulable matter of the blood; these fibres have been seen in the clot-cake which the blood, left to itself, never fails to form, and which seems to be really a sort of network made of small membranes which can be separated from the fluid part and can then be plainly seen."

But Haller did not admit that the fibres were really an element of the blood. He said:

> "If the authors wish us to understand that these fibres are in the blood as are the globules, they are certainly in error."

In support of his view he cited Borelli, the mathematician, who had been the first to refuse to admit "the fibres among the elements of the blood, as also Boerhave and other great men who have followed him," adding further:

> "If the authors wish to say that under certain circumstances fibres and flakes are born in the blood, he did not object thereto..."

but observed that these fibres and flakes seemed to have their birth rather in the lymph than in the red particles of the blood. In short, according to Haller, the blood contained nothing solid and figured but the globules in a liquid called lymph, adding that he had recommended as a good way of rendering the globules visible the addition of certain salts to the blood which increased the fluidity and the colour; "nitre being of all salts that which gives the best colour to the blood."

Haller, who derived the fibres of the clot from the lymph of the blood, was the precursor of the savants, who, like him, saw in the blood only globules in suspension in a liquid where everything else was supposed to be in a state of perfect solution.

The circumstances of the formation of the clot, its shape depending on that of the vessel in which it was formed, its progressive contraction and expulsion of the yellow serosity, thence called the serum, were all observed with attentive curiosity. The blood having finished its contraction, the washings in water which dissolved its colouring matter furnished the white matter which was called the fibrous portion of the blood, and, after the reform of chemical nomenclature, fibrin. The fibrin was finally isolated from the blood by whipping, before it coagulated. The great German physiologist, J. Muller, agreed with Haller. He wrote:

> "By liquor of the blood (liquor, lympha sanguinis) we mean the colourless liquid, such as exists before

coagulation, in which the blood globules swim ... it contains all that is really dissolved in the blood. At the moment of coagulation, the liquor separates itself from the fibrin which had before been dissolved";

and from his microscopic observations upon frog's blood, he thought that his researches

"proved that besides the albumen, the fibrin was dissolved in the liquor of the blood." [1]

Schultze gave the name of plasma to the lymph of Haller, which Muller had called *liquor saunguinis*.[2]

The conclusion of Muller was the more circumspect, seeing that it was a refutation or contradiction of another mode of considering it, already published. Hewson had expressed two views, one of which had agreed with that of Muller; the other was original. According to the former, the fibrin exists in the blood in a state of solution; according to the other, it exists in it in suspension in a state of fine granulations; he further admitted that the globules did not contain fibrin.

Milne-Edwards accepted the second opinion of Hewson, maintaining that the fibrin did not exist in solution in the blood, but in a finely divided state as a solid, under the form of fine granulations, which after the blood has been shed and left at rest, united together in the form of the fibres of the clot, or by whipping, to form fibrin.

Dumas, who, with Prevost of Geneva, had first admitted the globular origin of the fibrin to explain coagulation, afterwards accepted, to a certain extent, the opinion of Milne-Edwards.

It is important to explain the point of view of such a genius. He said:

" None of the properties of fibrin give us the means of explaining the state in which it exists in the blood. It has not been possible to bring back the fibrin to this condition by any known process. In fact, the blood contains the fibrin, both liquid and spontaneously coagulable...

Everything leads to the belief that this fibrin of the blood is not in solution in it, but that it exists there in a finely divided state, which it maintains so long as the liquid is in motion, but which, in the liquid at rest, stops all of a sudden as a consequence of the disposition of the fibrin particles to unite in a fibrous and membranous network." [3]

Later, he modified this view as follows:

> "Blood contains a quantity of spontaneously coagulable fibrin in suspension, or in a state so closely approaching solution, that it seems to be really dissolved in it; it is found there in a peculiar flowing state, analogous to that presented by starch mixed with water in an aqueous solution of starch."[4]

But neither of the views of Hewson, nor that of Milne-Edwards, nor that of the illustrious Dumas regarding the individual state of the fibrin in the blood, which, as will be seen, were the nearest to the truth, received much consideration, and were soon lost to view. Physiologists reverted more and more to the view of Haller adopted by Muller and Schultze.

The word "plasma" prevailed over lymph, and it was held that everything except the globules were in a state of complete solution in the blood. They came at last to believe that the blood did not contain fibrin even in a state of solution.

In short, the fibrin which was called the *corps de delit* of the coagulation of the blood was imagined in turn to be the same substance as albumen, and it was further imagined that:

> ... the albumen of the blood was none other than the fibrin combined with the alkali of the blood, only the part not so combined being coagulable;

> ... the plasma contained plasmine, which, when out of the vessels, transformed itself by spontaneous decomposition into concrete fibrin and into dissolved fibrin, called also metalbumen;

> ... the fibrin does not exist either in the blood or in the plasma; but that they contain, in solution, substances called fibrinogen and fibrinoplastin, respectively, which, outside of the vessels, under the influence of a ferment, produced the fibrin with an elimination of alkali, etc.

Chemists agreeing with Thenard came to look upon fibrin as an isolated animal matter, that is to say, a "proximate principle," according

INTRODUCTORY AND HISTORICAL

to the definition of Chevreul. Glénard, who paid a great deal of attention to the phenomenon of the coagulation of the blood and its causes, wrote upon the subject of fibrin as follows:

> "Science has not yet been able to establish the constitution of fibrin, of the "corps de delit" of coagulation; it is not known whether it be derived from albumen, or should be regarded as one of its stages; and the formula of this substance varies with each chemist; it is not known whether it is superfluous (recremen-titious) matter, or a product of excretion, a nutriment, or an organic waste." [5]

It is, therefore, a legitimate conclusion that after a century of hypothesis after hypothesis, we have gotten back to the point where Haller had left the question. Having neglected the conception of Milne-Edwards and of Dumas, as well as some researches which seemed to be approximately a verification thereof, it is no surprise that scientists who understand neither the real nature of fibrin nor its origin had recourse to occult causes for the explanation of the phenomenon of coagulation.

The celebrated English surgeon, Hunter, thought that:

> "blood coagulated by virtue of an impression, that is to say, that its fluidity being inopportune or no longer necessary in its state of rest after issuing from the vessels, coagulates in reply to the indispensable customs of solidity."

Also, he said that

> "the blood possesses in itself the force, by virtue whereof it acts in conformity with the stimulus of necessity, a necessity which is derived from the position in which it finds itself."

And Hunter wrote in the time of Haller.

Long after, Henle, having said that the cause of the coagulation of the blood directly after circulation ceased was unknown, added:

> "Coagulation is often regarded as the last act of life, as the death of the blood." [6]

This point of view, which was not that of Henle, has been lately revived and fitted into the system signified by the word plasma. In short, the following propositions can be collected from a work full of interesting observations on the coagulation of the blood:

> " The blood is endowed with a life of its own."
>
> " Coagulation is a synonym for the death of the blood."
>
> " By the fact of spontaneous coagulation, the plasma loses its chief property, that of living, and from the state of an organized humour becomes an inert aggregate of proximate principles."
>
> " Coagulation then is the disorganization of the plasma."
>
> "It is the fact of this organization which struggles for some minutes against the fatal influence upon the shed blood of contact with foreign bodies." [7]

Right here, before going further, will be the place to seek for the substance beneath the mask of words.

It is true that the author of the above propositions did not, like Hunter, invoke 'an impression' or 'the indispensable customs of solidity' nor 'the stimulus of necessity' to explain the phenomenon of the spontaneous coagulation of the blood, but has he escaped the shoals of 'occult causes'?

It is true that blood as it issues from a living body is alive. But is it not an 'explanation' by the occult causes to say that the blood coagulates because it dies?

But if the chief property of the plasma, an organized humour, is to live, is not the struggle of its organization against the fatal influence of contact, the loss of its life, also an 'explanation' by occult causes?

Also, the plasma, being an aqueous liquid in which the materials composing it cannot be other than proximate principles, are by the hypothesis and by definition in a state of perfect solution, is it not an explanation by occult causes to say that the cause of its spontaneous coagulation is its disorganization?

And what is the value of explanations by occult causes? Here is the answer given to this question by Newton:

> " To say that each species of things is endowed with a specific occult quality, by means whereof it has a certain power of action, and can produce sensible effects, is to say nothing at all."

Nevertheless, if in 1875 the author (Glénard) was reduced to the extremity of seeking an explanation of the phenomenon in considerations outside of anatomy, physiology and chemistry, it was because the then state of science did not offer anything more satisfactory. There are to be found in the transactions of the Academy of Sciences of the same year attempts at explanation which compared the so called coagulation of milk to that of the blood.

Still later, Frey, returning to the methods of Muller and of Haller, said:

> "Studied from the anatomical point of view the blood offers for our consideration a transparent colourless liquid, the plasma or liquor sanguinis, wherein float two kinds of cellular elements; the coloured cellules or red globules and the colourless cellules or lymphatic globules."

And as regards the fibrin, he says:

> "It is not known under what form it exists in the liquids of the organism before coagulation, and it is generally supposed to be a derivative of albumen." [8]

That amounts to saying that the red globules and the leucocyte are the only figured elements of the blood, and that the plasma holds the materials composing it in perfect solution, as Muller thought he had demonstrated for his liquor sanguinis, these materials being reducible from the organic point of view to albumen.

Further, Frey so thoroughly believed this that he said:

> "The rapid nutritive exchanges which are produced in the nutrient liquid of the organism hinder the formation of fibrin during life." [9]

All of which amounts to saying that at the moment of shedding the blood does not contain fibrin.

And here it may be observed that neither Haller nor Muller had any prejudices on the subject of the innate nature of the lymph or liquor of the blood. On the other hand, when plasma is made a synonym for 'liquor sanguinis' the question is prejudged, for the synonym *plasma* is attached to a particular conception of organization and of life, in conformity to the system which asserts that "life is a special form of the activity of matter," a system which differs greatly from the doctrines of Bichat, according to which life is not attached directly to matter, but to anatomical elements limited as to their form and structure. On this I shall insist further in explaining anatomically and physiologically the spontaneous coagulation of the blood.

But several years before Glénard and Frey wrote, Béchamp and Estor had demonstrated that the blood contains, besides the two species of globules, a third figured element, clearly determined in form and properties, by means whereof the phenomenon of coagulation could be explained without any recourse to occult causes.[10]

In his thesis Glénard referred to our researches in these words:

> " For reasons which we shall not fail to develop in a later work, we suppress a chapter having for title, *Theory of Béchamp and Estor on the Microzymas.*"

I do not know whether Glénard has anywhere developed his reasons for suppressing the above entitled chapter from his thesis. For my part I had the great sorrow of not being able to continue and complete with Estor the work we had commenced together. A separation which occurred in 1876, and then the so premature death of Estor, deprived me of my eminent collaborator and devoted friend; I had to pursue alone the complete solution of the problem. My latest researches have been carried on in the laboratory which Friedel provided me with at the Sorbonne.

The partial results of my researches have been described in notes which have appeared in various magazines; the last, in 1895, was in the form of a communication to the Congress of the French Association for the Advancement of Science held at Bordeaux; but several portions, and that especially which is the crown and keystone of the work, remained unpublished until the appearance of the present work.

The discovery of the third figured element of the blood was not made during the investigation of the phenomenon of the spontaneous coagulation of the blood; but Estor and I applied it according to the ideas then prevailing to the production of fibrin after phlebotomy, to explain the formation of the clot. When I resumed my study of fibrin from the point of view of blood-coagulation, I had already solved the problem of the coagulation of milk in a sense very different from received ideas, and this was long before the publication of the thesis of Glénard, who said:

INTRODUCTORY AND HISTORICAL

> "Not only are we ignorant of the first cause of coagulation, but we do not even know its proximate cause; we do not know whether this change in the state of the blood is a physical or chemical phenomenon; whether it is a crystallisation or a precipitation."

Unless I am much mistaken, that implies that the author doubted even what Haller, and later, Muller, Hewson, Milne-Edwards and Dumas held as certain, i.e., that the formation of the clot had the fibrin for its direct and near cause. As to the assertion that coagulation is a *variation of the state of the blood*, etc; it proves that its author did not know either the anatomical or chemical constitution of the blood any more than that of milk.

In our note of 1869, the microzymas of the blood were expressly mentioned as being the first cause of the production of fibrin and the proximate cause of coagulation. My new researches further demonstrated that the presence of the microzymas and that of the fibrin in the blood are correlative, the one presupposing the other; it was only necessary to explain this correlation to verify and complete the conception of Milne-Edwards developed by Dumas.

These new researches were allied to others, both older and newer, regarding the determination of the causes of the changes, reputed to be spontaneous, of organic matters, even of proximate principles in general, and specially of natural vegetable and animal matters, i.e:

1) The question of the origin of ferments and the physiological theory of fermentation.
2) The resolving in the negative the problem of the supposed spontaneous generation of ferments.
3) The origin of urea in the organism during the act of respiration.
4) The chemical constitution of albuminoid matters and the demonstration of the definite specificity of their chemical molecules.
5) The true theory of organization according to the doctrine of Bichat.

It is thus seen that the complete solution of the problem concerning the spontaneous coagulation of the blood necessitated the previous solution of several other problems very difficult of solution; they are given here nearly in their chronological order.

THE BLOOD AND ITS THIRD ELEMENT

1) The nature of fibrin, isolated from the clot, or obtained by whipping.
2) The real specific individuality of the albuminoid proximate principles.
3) The state of the fibrin in the blood at the moment of shedding.
4) The real structure of the red globules of the blood.
5) The real constitution of the blood at the moment of shedding.
6) The real chemical and physiological meaning of the coagulation of shed blood.

These will be the captions of the following chapters.

After the developments which are to follow, it will be possible to understand that what is called the phenomenon of the spontaneous coagulation of blood is not at all a coagulation of the blood itself, but of that of a portion of its third anatomical element.

It will then clearly appear that that which is improperly called a coagulation is only the first phase of a much more complete alteration of the blood involving the destruction of its blood globules and other changes, even that of its red colouring matter; and further, that this spontaneous alteration of the blood is but a special case of a very general phenomenon, that of the spontaneous alterability of all animal matter, solid or humoural, abstracted from an animal, whether living or dead; an alterability, physiologically spontaneous, necessary, drawing with it the destruction even of the cellular anatomical elements themselves, as the consequence of phenomena of fermentations of a special kind, whereof *the microzymas of these matters are the principal agents.*

INTRODUCTORY AND HISTORICAL

NOTES

1. J. Muller, *Manual de physiologie*. Littre's ed., Vol. 1, p.95 (1851).
2. Henle, *Anatomié générale*. Trans. (Fr.) by Jourdan, Vol. 1, p.444.
3. Dumas, *Traité de chimie appliqué* aux arts, Vol. VII, p.451 (1844).
4. *ibid* Vol. VIII, p.478 (1846).
5. Glénard, *These sur la coagulation spontanée de sang*. (1875).

    [It is worthwhile to note the nonsensical invention of names for imaginary substances—which no one has ever seen or will see, and to contrast the nonsensical 'reasonings' of some men of science, with the beautiful simplicity and *reality* of the exposition given in this work by Professor Béchamp.—Trans.]

6. Henle, *Anat. gén.* trans. from the German into French by Jourdan, Vol. 1, p.39.
7. Thesis above sited, pp.63-65.
8. Frey, *Traité d'histologie et d'hist-chimie*. Fr. tr. P. Spielmann, p.120 (1877).
9. *ibid*, pp.14-15
10. *C. R.*, Vol. LXIX, p.713 (1869).

CHAPTER ONE

On the nature of fibrin isolated from the clot or obtained by whipping the blood.

The blood fibrin.

Fibrinous microzymas.

Fibrin and oxygenated water.

The ferment of fibrin.

Guy-Lussac and Thénard analysed fibrin as they had analysed albumen, casein and gelatine. Thénard said that fibrin was an isolated animal matter; Chevreul said that it was an animal proximate principle and was greatly surprised, after he had discovered oxygenated water, to find that fibrin decomposed it and disengaged the oxygen, as did organic tissues, for example the liver. He even thought that fibrin was the only proximate principle of its kind endowed with this property.[1]

This fact in the history of fibrin is important; first, because it is the pivot on which turns the demonstration that this substance, reputed a proximate principle, is of the same order as the substance of the bodies which Chevreul called organic bodies; secondly, because, although neglected by physiologists and chemists, it enabled me to place beyond doubt the existence of a third anatomical element of the blood.

I did not set out with the idea of proving that fibrin is a substance of the same order as the organic tissues. Like everybody else, I regarded it as a proximate principle; I had even maintained its specificity against the chemists who contended that it was only coagulated albumen.[2]

Preliminary steps to the discovery of the real nature of fibrin and of the third anatomical element of the blood.

The ancients regarded it as a positive fact that all animal or vegetable matter was spontaneously alterable while putrefying or fermenting. In the last century (the 18th) the chemist Macquer established the

CHAPTER ONE

conditions for these changes; the presence of water, the contact of air and a certain amount of heat.

Long after, when in 1837 Cagniard de la Tour regarded beer yeast as being organized and living, and fermentation as an effect of vegetation, Schwann, generalizing the new conception, endeavoured to show that no organic matter was spontaneously alterable; that the alteration was caused by the presence of organized living things, microscopic cryptograms, vibrioniens; that is to say, ferments, the origin of which, reviving the old hypothesis of Spallanzanil, he ascribed to the germs of the air.

But, notwithstanding many important verifications, the opinion of Schwann did not prevail; the presence of living products in matter undergoing change was conceded, but while some maintained that the alteration preceded the appearance of the organized products, whatever might be their origin, others, admitting the theory of Cagniard, insisted that the living things, the ferments, were the fruit of spontaneous generation.

Schwann's point of view and the hypothesis of germs of the air were so completely abandoned that in 1884 it was admitted as a fact that even cane sugar in watery solution altered spontaneously at the ordinary temperature of the air, becoming what was called invert sugar, grape sugar.

Was that true? The inversion of cane sugar was the result of a chemical reaction of reduction by hydration which was produced, as Biot had observed, under the influence of strong acids. Could it be effected by water only, at ordinary temperature with the aid of time alone? I wanted to know what to believe, and I instituted experiments which commenced in 1854 and would continue until 1887.

Several consequences of the greatest importance resulted. Among them was the first experimental verification of the hypothesis of germs of the air which Schwann, following Spallanzanil, had invoked against spontaneous generation.

In short, I demonstrated:

1) That a watery solution of cane sugar remains unaltered indefinitely, at ordinary temperature, under either of the two following conditions:

a) absolute protection from access of air; or

b) in contact with a limited volume of air, to which have been added certain salts or a suitable (small) quantity of creosote; for instance one to two drops per 100cc.

2) That the same solution, either pure or with the addition of certain other salts in contact with the same volume of air, permitted the appearance of cryptogamic products, moulds, etc, at the same time that the inversion of the sugar was effected.

3) That the moulds are actually the agents, the ferments of the inversion, by secreting the necessary zymas or soluble ferment.

4) That creosote, which hinders the birth of the moulds, etc, does not prevent developed moulds from effecting the inversion.

As it is evident that the water and sugar of the solution cannot of themselves give birth to those cryptogamic productions which invert the cane sugar, nor to anything whatever organized and living, the conclusion is inevitable that these experiments verified the hypothesis of the existence of germs in the air.[3]

Cane sugar being a proximate principle, the experiment constituted also the first demonstration that organic matters exist which are unalterable under the conditions specified by Macquer.

To be applicable generally, it was necessary to prove that what was true for cane sugar was also true for any proximate principle, even for albumin, supposed to be so readily alterable that Colin believed it could spontaneously become an alcoholic ferment.[4] But there are solutions of proximate principles, even of their mixtures, which contain some albuminoid substance like the solutions of cane sugar; these solutions, with a very small quantity of creosote added, are preserved, although in contact with a limited quantity of air, so that nothing organized makes its appearance; no fermentation, no putrefaction takes place. But if among the materials of the mixture there are some which are directly oxidizable by the oxygen of the air, creosote will not hinder the oxidation.

Let us bear in mind this fact which has been experimentally verified in every imaginable case:

The solutions of isolated proximate principles, or mixtures of them, even albuminoid ones, first creosoted with a suitable (small) dose, and exposed to the contact of a limited quantity of ordinary air, allow nothing living to appear and remain unaltered except in cases where the mixture contains some directly oxidizable principle.

In these kinds of experiments the creosote acts either by rendering the medium sterile for the germs or directly upon them, so as to prevent their development.

Organic matters reduced to proximate principles are unalterable under the conditions specified by Macquer, when the influence of the germs of the air is prevented by creosote; then they are so naturally. But Macquer did not take into account proximate principles, of which he had no idea. He really referred only to natural vegetable and animal matters, that which Thénard calls organic tissues and Chevreul organic bodies.

But among the animal matters on which Macquer experimented was milk, which he regarded as an animal emulsion, and held to be alterable of itself. Much later, Donne (an expert micrographist) and most chemists regarded milk as a solution of milk sugar, casein and of mineral salts, holding an emulsion of butter in solution. Everybody then thought that milk was a pure mixture of proximate principles.[5]

Such a mixture properly creosoted and in contact with a limited volume of air ought to remain unaltered indefinitely. But it was found to be otherwise. Cow's milk sufficiently creosoted at the time of milking, preserved from contact with air or in contact with only a limited amount of air, neither sours nor clots in the ordinary way. The creosote only delays the souring and the consecutive formation of the clot. But it was found that at the moment the milk became clotted, even when the phenomenon takes place in full contact with air, and with or without the addition of creosote, none of the cryptogamic productions could be found which Schwann's experiments led one to look for in it.

But the souring and the clot, I do not speak of the coagulation of the milk, is only the first phase in the phenomenon of alteration. The second phase, in spite of the addition of creosote, was characterized by the constant appearance of vibrios or of bacteria. Milk then does not act as would a simple mixture of proximate principles.

THE BLOOD AND ITS THIRD ELEMENT

These experiments and observations, which date from before 1858, were not published until 1873.[6] They had greatly surprised me.

Milk, then, was not what it had been supposed to be. There exist in it organic matters alterable without the aid of germs of the air, and Macquer was justified in declaring them to be spontaneously alterable. And since, notwithstanding the creosote the milk already altered, soured and clotted, permitted the appearance of vibrios in its substance, if these vibrios were not the products of spontaneous generation, to what did they owe their birth?

The answer lay in experiments contemporary with those upon the calcareous rocks which will be studied in the last chapter of this work, as well as those which led to the discovery of the new category of living organized productions to which I have given the name of microzymas, because of their functions as ferments and of their extreme minuteness.[7] It is then the microzymas of the milk itself which are the agents of its alteration and which subsequently become vibrios by evolution.

The method which led to these results, and demonstrated the close relationship between the geological ferments and the anatomical and physiological ferments of present living animals, and which at the same time answered in the negative the question of the spontaneous generation of organized ferments, is the same which has permitted the demonstration of the inherent inalterability of proximate principles, and has also verified the old hypothesis of germs of the air which had been neglected. Thanks to it, it has been possible to explain anatomically and physiologically the phenomena of coagulation and the other spontaneous changes of the blood.

This method had its origin in experiments on the inversion of cane sugar, supposed to be spontaneous, and in those relating to the changes which occurred in milk, which made conspicuous the principle, obtained by experiment, that creosote absolutely prevents the alteration of immediate principles by preventing the development of all living organized products, even in contact with a limited quantity of ordinary air, while the same doses under the same conditions did not prevent change in natural animal matters, tissues and humours, even permitting them to give birth to vibrios or to bacteria.

It is important to bear in mind that the new method (of experiment)

CHAPTER ONE

enabled us to distinguish organic matters composed only of proximate principles, from natural vegetable and animal matters, that is to say, from organic bodies properly so-called; in short, to distinguish organic matter in the chemical meaning from that which, like milk, is organic matter in the anatomical and physiological meaning. The ferments which change the former, the organic matters in a chemical sense, that is to say, proximate principles, have for origin the germs of the air, while the ferments which change the second, that is to say, the natural organic matters, are the microzymas of their own substance, which are inherent in them as anatomical elements.

In fact, the phenomenon of the birth of the vibrios in the spontaneously altered milk, if indeed they were the result of the evolution of the microzymas of the milk, ought not to be an isolated fact, but a particular case of a general phenomenon, proper to all organic bodies, so that the fact of the birth of vibrioniens in an organic body, humour or tissue should be considered as evidence of the existence of microzymas in this tissue or this humour, even though the microscope has failed to reveal them.

Experimentation has confirmed in every sense these consequences of the application of the new method of investigation to the study of the phenomenon of the spontaneous change of milk. The matter of all the tissues and humours, of all organic bodies, from the highest to the lowest—as, for instance, beer yeast and the mother of vinegar—may give birth to vibrioniens in conditions similar to those in which they are produced in the milk, or in which such conditions can be realized, if it be necessary to encourage otherwise the evolution of their microzymas. And when the phenomenon of spontaneous alteration of such matter is recorded, the matter being protected from germs of the air, without the appearance of vibrioniens, it may confidently be affirmed that microzymas were present and were the agents of the change.

The following is the application of the method to fibrin, regarded as an organic body.

Demonstration that fibrin is not a proximate principle, but a false membrane of microzymas. The birth of bacteria in the fibrin.

The fibrin is produced mechanically from the blood by whipping the latter; being regarded as an organic body, it ought like milk to contain microzymas capable of undergoing vibrionian evolution.

To demonstrate this, Estor and I employed a modification of the method which had been used to demonstrate the microzymas of the chalk and of muscle flesh.

The modification consisted of preparing a starch of the fecula of potatoes, boiling it for a long time, creosoting it while boiling, and introducing into it the solid substance to be studied at the moment it was extracted from the creosoted water, into which it had been immersed to protect it from the influence of germs of the air.

The experiment was as follows:

Fibrin was obtained by whipping under the following conditions; at the moment of the venesection, creosote was added to the blood, and at the same time it was whipped with a bundle of metallic wires which had just been washed in boiling creosoted water; then the fibrin was washed in creosoted water.

Into 100g of creosoted starch, 15g of freshly prepared humid fibrin was introduced, and the flask sealed and placed in an oven heated to 30° to 40°C (86° to 104°F). The starch, exactly as happened with muscle issue, became liquefied by degrees, and after a time the presence of bacteria in the mixture was evident; but it was observed that the liquefaction of the starch generally preceded the appearance of the bacteria.

The foregoing is a general view of the phenomena, but differences were observed in its manifestations, according to the species and age of the animal, as well as the source of the blood. Generally the fibrin of young animals disintegrates in fluid starch, while the bacteria develop. The duration of the liquefaction of the starch is also variable.

It is known that boiled milk clots, which means that the microzymas are not killed at the temperature of boiling; on the other hand, to prevent the chalk from liquefying starch,[8] I was obliged to heat it (moist) to more than 200°C (392°F). The microzymas of the fibrin resist up to 100°C (212°F). The fibrin was boiled for several minutes in distilled

CHAPTER ONE

water before it was placed in the starch. In this case the liquefaction is still further delayed, and even ceases to be produced if the boiling of the fibrin is too prolonged; but the bacteria appear, none the less, and these bacteria always present the same morphological characteristics.

To complete the demonstration, it should be added, and M. Estor was witness of the fact, that fibrin, exactly as was the case with the mother of vinegar (a sort of vegetable membrane of visible microzymas) and under like conditions, can produce lactic and butyric fermentation, a fact which will be further considered hereafter.

Such was the experiment which concluded that fibrin, like milk, like flesh, like the tissue of liver, etc, contains microzymas, since, like them, it gives birth to bacteria without the aid of germs of the air.

Under the conditions of the experiment these microzymas could only have come from the blood. Efforts were then made to find them in the blood itself at the moment of the venesection. This was a delicate investigation and will be referred to hereafter, as it is allied to the whole subject of this work.

Fibrin, whether obtained by whipping or by washing the clot (we will see presently wherein these two preparations differ), is not a proximate principle, but constitutes a membrane or fibre composed of microzymas. In short, it is not organic matter in the chemical sense, but an organic body in the anatomical and physiological sense.

Nevertheless, this demonstration that fibrin contains microzymas is indirect; and one might still contend that the bacteria has been born spontaneously, in the mixture of fibrin with starch. In any case, it left unsettled the nature of the substance which, in the false membrane, is like an intermicrozymian gangue, and also the question of the quantitative relation between the microzymas and this substance. It was therefore very desirable to obtain these microzymas isolated, as Estor and I had isolated those of the liver.

The fibrinous microzymas and their properties compared with those of the fibrin.

The question of the solubility of fibrin in dilute hydrochloric acid had been long discussed, and it occurred to me that I might find there a means whereby the microzymas might be isolated, as they ought to be insoluble in it as were those of the chalk. On going over the history

75

of experiments on the fibrin, I found many experiments and observations relating to fibrin deserving attention and which had been long neglected.

Thénard had already described the action of dilute hydrochloric acid upon the fibrin of the blood and the formation of hydrochlorates of this substance, one of which, the gelatinous, was soluble in tepid water.[9] Long afterwards, Bouchardt called attention to the fact that Chevreul had demonstrated that fibrin always contained fat and he asked himself if even deprived of fatty bodies it would be a pure proximate principle.

To demonstrate that that which had been generally admitted was not well founded, he made the following experiment:

He heated fresh, moist fibrin with ten times its weight of very dilute hydrochloric acid (1 to 2000) and observed that it swelled up, and by a prolonged maceration was at last dissolved, but that there always remained manifest a portion of a product which was not attacked by an excess of this very dilute acid employed as a solvent. Bouchardt called the undissolved part *epidermose*, and the dissolved part albuminose.[10] It was from this experiment that Bouchardt justly concluded that the fibrin is decidedly not a proximate principle.

The part of the fibrin that Bouchardt regarded under the name of epidermose, as one of the two proximate principles constituting fibrin, was precisely the microzymas which I proposed to isolate. In consequence of the viscosity of the hydrochloric solution, essentially capable of change as Bouchardt had said, and of the slowness of filtration, he had not determined the quantity of the epidermose.

By employing a less diluted hydrochloric acid (1 to 3cc of fuming acid per 1000cc) the viscosity was diminished, and on adding to it 2-3 drops of phenol per 100cc the suspected alteration was absolutely prevented, and the deposition of the insoluble part could be awaited. Even then it requires from ten to twelve days to filter a litre of the liquid. For quantitative experiments it must be left at rest and the deposit must not be turned out upon the filter until the filtration of the supernatant liquid is completed.

The grayish brown mass retained on the filter was resolved under the microscope into exceedingly fine molecular granulations, which are the microzymas, and some shapeless fragments proceeding

CHAPTER ONE

doubtless from the blood globules destroyed during the preparation of the fibrin. To procure these molecular granulations as free as possible from foreign fragments the mass when removed from the filter is steeped in hydrochloric acid (1 in 1000), and the liquor, creosoted or carbolised, is passed through a fine linen mesh and left to deposit.

The deposit is collected on a filter of very fine mesh, is successively washed with water to remove every trace of acid, and finally washed with ether, slightly alcoholised, to remove the fat. The matter when removed from the filter is dried in a dry vacuum, is agglomerated, and brownish in colour.

The moist fibrinous microzymas, completely drained, are composed (in hundredths) as follows:

Organic matter, chiefly albuminoid	13.553
Mineral matter	0.384
Water (by difference)	86.063
Total	100.000

Like all organized beings, they contain mineral matter and much fixed water. Their organic matter is chiefly albuminoid; in fact, dry, it dissolves in fuming hydrochloric acid, developing a violet colour when hot; and if water is added to the hydrochloric acid, a white precipitate of albuminoid matter is obtained.

The minuteness of these humid microzymas, swollen with water, is extreme. Under the microscope they appear to be spherical in form, animated with the brownian movements, the diameter whereof hardly attains 0.0005 (half a thousandth) of a millimetre.

Their quantity is very small. From some determinations, unavoidably somewhat uncertain, I estimated that the humid, drained fibrin of the general blood of an ox yields about one thousandth of its weight in microzymas dried at 100°.

Taking this figure of 1/1000 as the best approximation, and considering that 1,000g of drained fibrin contains 193g of fibrin dried at 100°, it is evident that the weight of the dried microzymas is 1/193 of the dry fibrin; in short 100 parts of fibrin dried at 100° contain 0.518 parts of microzymas dried at the same temperature.

This quantity appears to be very small, and one might think that in the blood it might be neglected, and that consequently the microzymas take no part in the phenomena studied. It is not so, for we shall see that they are anatomical elements and physiological agents of rare energy; and that if it was interesting to weigh them, it is still more so to count them.

Let us first show that in the fibrin they are at the same time that which liquefies starch and from which bacteria are derived, that which decomposes oxygenated water, and that which determines its apparent solution in dilute hydrochloric acid.

I. The fibrinous microzymas liquefy starch and then become bacteria

The microzymas of 60g of fibrin obtained from the blood of an ox or of a dog, fresh, still humid, well washed so as to remove every trace of acid, are sufficient to liquefy 50g of potato starch at 45° to 50°C (113 to 122°F). The liquefaction is completed in 16 hours; if the reaction is prolonged, Fehling's reagent is reduced. Other things equal, the liquefaction is more rapid with the microzymas of the fibrin of a dog. Finally bacteria appear, while another fermentation begins and the liquid becomes acid.

To estimate the influence of the concentration of the acid in the extraction of the microzymas, the fibrin in another operation was treated with hydrochloric acid, 3 to 1,000. The microzymas did not lose any of their activity.

2. The microzymas of fibrin decompose oxygenated water

The humid microzymas, crude, or with the fat removed by ether, as well as those that have been dried in a dry vacuum, decompose oxygenated water, setting the oxygen free, but with much greater energy than the fibrin from which they had been obtained; showing therein an energy hardly less than bioxide of manganese. I ascertained that the microzymas of the fibrin of the blood of all the animals examined by me acted in like manner.

Later the theory of these facts will be explained, but to anticipate the objection regarding the germs of the air I call attention to the four facts following:

a) Fibrinous microzymas which have liquefied starch are still able to decompose oxygenated water.

CHAPTER ONE

b) Fibrinous microzymas which have exhausted their decomposing action upon oxygenated water can no longer liquefy starch and do not develop into bacteria.

c) Fibrinous microzymas which have been subjected to boiling at 100°C (212°F) do not liquefy starch and do not decompose oxygenated water.

d) Fibrinous microzymas lose the property of decomposing oxygenated water with the passage of time.

But fibrinous microzymas washed in ether, so as to remove their fat, dried in vacuo and protected from contact with the air in a closed tube, preserve for a long time the property of decomposing oxygenated water, but lose by degrees their energy; after ten years they had lost it altogether, without having appreciably lost weight.

Here was another essential property of the microzymas which I formulated!

3. Fibrin owes to its living microzymas the faculty of being dissolved in very dilute hydrochloric acid

Bouchardt, following Thénard, observed that before dissolving in dilute acid, fibrin swelled up in a translucent, colourless gelatinous mass[11] and that solution was attained only after prolonged maceration. The progress of solution is so slow that Liebig, for a long time, held that fibrin was insoluble in dilute hydrochloric acid; and we shall see that it was upon this remark that he founded the distinction between muscular fibrin (masculine or syntonine) and blood fibrin.

Dumas, on the other hand, verified the fact of solubility and showed that at the temperature of 400°C (104°F) solution was more rapid. According to Dumas the phenomenon is a function of time and temperature. I shall prove that it is at the same time especially a function of the activity of the microzymas.

First, remember that creosote or phenic acid delays the souring and coagulation of milk as well as the vibrionian evolution of its microzymas. Phenol similarly retards the supposed solution of fibrin in very dilute hydrochloric acid. The following demonstrates the fact:

A mass of 600g of fresh and humid fibrin of ox's blood is divided into four equal parts; A, B, C, and D, each of 150g, which are treated in flasks of like capacity in the following manner:

A: 2,000cc of hydrochloric acid, 2 to 1,000;
B: the same volume of acid and 40 drops of phenol;
C: the same volume of acid and 60 drops of phenol;
D: 2,000cc of boiling distilled water.

The ebulition is maintained at 100°C for two minutes, and left to cool. 4cc of fuming hydrochloric acid is added, so that it was also diluted to 2 to 1,000.

The four flasks were covered and placed in the same enclosure. The temperature was kept at 24° to 28°C (75.2° to 82.4°F).

In A, B, and C, the fibrin swelled into a gelatinous mass. In D, the fibrin remained a dull white, without becoming gelatinous.

In A, the gelatinous mass was dissolved in three days.

In B, the solution was effected in four days.

In C, the solution was effected in six days.

In D, the unswollen fibre remained a dull white; there was no change at the end of a fortnight, though with free access of air.

The phenomenon at the same temperature is then a function of time; it must also be so of the microzymas, since phenic acid retards it the more, the greater the dose, even as it delayed the coagulation of milk, and finally boiling for a sufficient time prevented it entirely, as it had prevented the fibrin and the fibrinous microzymas from liquefying starch and from decomposing oxygenated water.

The function ascribed to the microzymas will be made still more clear when it is shown that that which is called the dissolving of fibrin is really the result of a reaction of a profound transformation undergone by that part of the fibrin which is in solution. The theory of the phenomenon will also be explained presently; for now we will confine ourselves to saying that in the order of the ideas of these experiments the supposed solution in very dilute hydrochloric acid is, at bottom, only a mode of spontaneous alteration of the fibrin under special conditions.

We have now to consider the normal method of its spontaneous alteration.

CHAPTER ONE

4. Fibrin changes spontaneously without undergoing fetid putrefaction

Guy-Lussac had observed that fresh fibrin, in an open flask, in contact with water that was renewed from time to time, putrefied and disappeared almost wholly, leaving only an insignificant insoluble residue. At the time this observation was made, it was believed that albuminoid proximate principles, as well as others, were spontaneously alterable. This was before the experiments of Schwann regarding the influence of the germs of the air.

On the study of this change being again taken up to determine its products, among those which are dissolved there was observed an albuminoid matter coagulable by heat, which was taken for albumin, also leucine, valeric acid, butyric acid, hydrosulphate of ammonia, etc. In reality, in the experiment of Guy-Lussac, the alteration was a complex phenomenon, in which the ferments born of the germs of the air take part and are the agents of the fetid putrefaction.

If the influence of these germs be annulled, the result is different. A mass of fresh fibrin, prepared with the usual care, was immediately immersed in distilled water (first carbolised by 3-4 drops per 100cc), so that it was covered with a bed of liquid.

Under these conditions, after five to six weeks, at a temperature ranging from $15°$ to $25°C$ ($59°$ to $77°F$) the fibrin had disappeared. In its place were a clear transparent liquid and a considerable precipitate. There was no odour except that of the carbolic acid, and no vibrios either in the liquor or in the precipitate.

The alteration, then, had taken place without any fetid putrefaction. What was its nature? In Chapter 2, the nature of these dissolved bodies will be compared with those of the changed fibrin in dilute hydrochloric acid. Let us see of what the precipitate consisted.

The molecular granulations of the change without fetid putrefaction of the fibrin.

In the precipitate, which is greater than the precipitate of microzymas after the disappearance of the fibrin in dilute hydrochloric acid, the microscope shows us a very great number of very small spherical molecular granulations, much more bulky than the fibrinous microzymas, and some shapeless remains, probably of fibrin or of the envelopes of blood globules.

To procure these molecular granulations in a pure form, the precipitate, which is thick, is steeped in slightly carbolised water, then passed through a close-meshed silk cloth, purified again by levigation, collected on a filter, to be there again washed with water and finally washed with ether slightly charged with alcohol to remove the fat, and then again with water.

In this condition the molecular granulations preserve their form; they decompose oxygenated water, liquefy starch and again decompose oxygenated water after having effected this liquefication; in short, they posses the properties of fibrin and of its microzymas, but they are neither fibrin nor its microzymas.

In fact, these molecular granulations, the insoluble remains of the disappeared fibrin, treated with hydrochloric acid (2 in 1,000) dissolve much more rapidly than the fibrin, leaving undissolved microzymas identical with, as slender as, and endowed with the same properties as, those of fresh fibrin.

This last observation is important. It is a consequence of the fact that fibrin, under the conditions of the experiment, alters spontaneously without putrefaction, without vibrios, leaving a residuum of molecular granulations which contain microzymas identical with those obtained from fibrin treated with dilute hydrochloric acid. It is explicable only in one way.

As milk, treated with a sufficient dose of phenic acid, becomes changed otherwise than milk not so treated or only slightly so, without the microzymas becoming vibrios, so the microzymas of the fibrin have transformed, in a certain way, the intermicrozymian substance which is in it, as gangue, without undergoing vibrionian evolution, but remaining enveloped as in an atmosphere of albuminoid matter insoluble in water, but easily soluble in very dilute hydrochloric acid, the microzymas being set free.

The great importance of taking these molecular granulations into consideration will be seen when studying in the third chapter the state of fibrin in the blood. Meanwhile, the fact that the fibrin changes spontaneously in carbolised water, that is to say, without the aid of germs in the air, is a fresh proof that fibrin is not a proximate principle.

CHAPTER ONE

In the next chapter, we shall investigate the nature of the albuminoid matters of the spontaneous alteration of fibrin in carbolised water, and compare it with that of the change which takes place under the influence of hydrochloric acid.

Meanwhile, such are the proofs, all agreeing, founded on the new method of investigation, to the effect that fibrin, like milk, the liver, etc, is neither a proximate principle nor a compound of such principles, but that like them it is an organic body containing special microzymas. Further, these living microzymas are what, in the fibrin, liquefy starch and can become vibrionen by evolution. They decompose oxygenated water, and determine the change of this fibrin either in very dilute hydrochloric acid or in carbolised water.

To complete the history of the microzymas of the fibrin, we must try to discover by what mechanism they decompose oxygenated water and liquefy starch, either isolated or in fibrin; and how it is that they are the agents which determine the spontaneous alteration of fibrin, both in very dilute hydrochloric acid and in carbolised water.

Theory of the decomposition of oxygenated water by fibrin and by the fibrinous microzymas.

I stated at the commencement of this chapter that Thénard, having discovered that organic tissues (for example, the liver) decompose oxygenated water, thought that fibrin decomposed it through being a proximate principle and was the only substance of this order that did so.

But what is really the nature of the phenomenon of this decomposition?

Thénard said that fibrin and organic tissues "decompose oxygenated water in the same manner as metals (platinum, for instance) without giving up any of their principles, without absorbing the smallest quantity of oxygen, without undergoing the least visible change."

In short, that oxygenated water is decomposed by fibrin owing to what has since been called "action through presence," or "catalytic action of contact," with such as metals of the bioxide of manganese. Such was the state of science a few years ago and is so, perhaps, today.

It was necessary for a more exact knowledge of the blood and of organization in general to establish the meaning of this, both as to fact and as to principle; the more so that they were advanced by Thénard himself as a possible explanation of the phenomenon of fermentation, and were the point of departure of the hypothesis called actions of presence, of catalytic content, which have been the cause of the true theory of fermentation being so much misunderstood.

In reality, the decomposition of oxygenated water by fibrin, with disengagement of oxygen, is the result not of an action by presence merely, as with the bioxide of manganese, but of a chemical reaction, as is evident from the following experiments:

30g of fibrin of fresh ox blood, containing 3.79g of matter dried at 100°C, have successively decomposed three times 60cc of oxygenated water at 10.5 volumes of oxygen.

At the second and third addition, the disengagement became gradually slower, so that at the third, after twenty four hours, no more gas was given off, although the oxygenated water was not all decomposed.

Altogether 1,600cc of oxygen were set free from 1,890cc, which the 180cc of oxygenated water employed contained. It is evident that if the fibrin had given up nothing, if there had not been some reaction, the successive liquors resulting from the action of the oxygenated water ought not to contain any organic matter. But these liquors on being evaporated left a combustible residue, whose weight, deducting the ashes, was 0.16g dried at 100°; i.e. 0.533 for 100 of humid fibrin or 2.76 per cent of fibrin dried at 100°C.

The fibrinous microzymas also yield up somewhat of their substance on decomposing oxygenated water. Six grams of these microzymas, fresh, humid, containing 0.84g of matter dried at 100°, having exhausted their decomposing action, the evaporating liquors have left as residue, dried at 100°, 0.86g of organic combustible matter, deducting the ashes; that is to say, 1 for 100 of humid matter, or 7.55 per cent of the weight of the dried microzymas.

Fibrin and its microzymas, then, do not decompose oxygenated water in the same manner as do platinum or the bioxide of manganese, since they both give up part of their substance which is found transformed in solution in the oxygenated water.

If Thénard thought that fibrin gave up nothing it was because, on the one hand, he took into account only the disengaged oxygen, which

CHAPTER ONE

seemed to him the whole of that which the oxygenated water could furnish, that which had been absorbed being very minute, and, on the other hand, the fibrin seemed to him not to have undergone any change. But the change really has been great, since that which remains has no longer any action on oxygenated water, does not liquefy starch, and does not yield bacteria.

These remarks apply to the microzymas which are recovered, and are similar morphologically to what they had been before being treated, but do not now liquefy starch nor become bacteria by evolution.

It is then a fact that decomposition with oxygen set free from oxygenated water by fibrin or by its isolated microzymas is correlative with a chemical reaction, with a change in the property of the substance which has exhausted its decomposing activity.

And on comparing, in hundredths, the quantities (of the products of the reaction) which are dissolved, of the fibrin and of the isolated microzymas, it is found that the latter furnish much more than the former. They furnish much more, even if we consider only the quantity of microzymas contained in the fibrin used, i.e. 0.0335g for 60g of humid fibrin or 5.79g of that dried at 100°C.

In fact, if the dried fibrin yields or contains 2.76, calculating that which its microzymas would give by comparison with what is given by the isolated microzymas, it is found to be 4 per cent instead of 7 per cent, which is that given by these latter.

I do not lay much stress on this difference because it may be due partly to the difficulties and uncertainties attendant upon the weighing. But none the less, by means of these comparisons, it is clear that the microzymas, whether isolated or not, give up more than does the fibrin, which tends to show that the intermicrozymian gangue of the fibrin does not exert any decomposing action upon oxygenated water, as will be presently demonstrated. Anyway, it is evident that some substance belonging to the organization of the microzyma—probably a proximate principle—is yielded and transformed, and it is not the entire microzyma which is the agent of the decomposition, since the greatest part of its mass remains, preserving its form.

But what is this substance? Without being able to define it exactly, we shall see that it is essentially albuminoid. Whatever it may be, it is important to know that it only effects the decomposition under certain conditions.

For instance, oxygenated water, which contains a free acid, is not decomposed by either fibrin or by the fibrinous microzymas, and reciprocally, the fibrin dissolved by the hydrochloric acid under the same conditions as the experiment of Bouchardt, in which the acid is very dilute, and containing microzymas, decomposes it only when it is neutral.

But albuminoid matters combine with several acids; it is without doubt a hydrochlorate, a sulphate, etc, of this substance, whatever it may be, which is not changed by oxygenated water, and oxygenated water which is not decomposed by it. In illustration of this the following very interesting case of the influence of a special acid is given:

Liebig observed that fibrin steeped in a very dilute solution of hydrochloric acid did not decompose oxygenated water. The observation was true but incomplete, as the influence of this acid is only temporary. In fact, if the quantity of oxygenated water is sufficient, the liberation of oxygen recommences after a period of time which is longer the greater the quantity of hydrocyanic acid. The decomposition recommences because the oxygenated water destroys the hydrocyanic acid by a phenomenon of oxidation without liberation of oxygen.[12]

Theory of the liquefication of fecula starch by fibrin and by the fibrinous microzymas.

Fibrin and its microzymas are insoluble in water; on the other hand, Payen demonstrated that fecula exists in a special condition of hydration and of swelling, in which it is similarly insoluble.

How then can these insoluble bodies act upon one another, the one, fecula, dissolving, while the other, the fibrin or the microzymas, remain insoluble? The explanation is the same as was given for the inversion of cane sugar by moulds, born of the germs of the air in its aqueous solution, but which are insoluble, as are the microzymas.

I have proved directly that these moulds, born of other organized ferments, and other microzymas, produce in themselves and secrete soluble products of an albuminoid nature, which are of the same order as those called soluble ferments and which were confused in the same category with organized insoluble ferments. Having thus established the anatomical origin of soluble ferments, to mark the union of

CHAPTER ONE

dependence between the product and the producer, I gave the name of zymas to what had been termed soluble ferment.

This established, as the microzymas[13] of sprouted barley produce diastase or hordeozymas, as the pancreas or its microzymas produce pancreazymas, which liquefy and saccharify starch, so the fibrinous microzymas produce the zymas which effect its liquefication.

And since every zymas is of the albuminoid order, as the fibrinous microzymas which have exhausted their decomposing action upon oxygenated water do not liquefy starch, we can say that the substance which in the fibrinous microzymas is given up and transformed by oxygenated water is precisely this zymas, an albuminoid substance which liquefies the starch of fecula.

Theory of the spontaneous alteration of fibrin, whether in very dilute hydrochloric acid or in carbolised water.

The two constitutive parts of fibrin are equally insoluble in water and in very dilute hydrochloric acid. As for the liquefication of starch by them, the same question arises: how can these two insoluble bodies act upon one another, the microzymas remaining insoluble, while the albuminoid matter enters into solution?

The answer is the same. In the same manner that fecula is made soluble and transformed by the zymas which the microzymas secrete, so the albuminoid matter is dissolved by this zymas while being transformed.

The explanation of the phenomenon is thus very simple. Only in the case in which very dilute hydrochloric acid intervenes does the transforming chemical action of the zymas secreted by the microzymas act upon the insoluble combination, which the albuminoid matter makes, at first with the hydrochloric acid; while in the carbolised water it acts directly on the insoluble albuminoid matter as on the amylaceous matter of the starch.

By the action of the zymas being exercised on the one hand on the hydrochloric combination of the albuminoid and on the other on this matter itself, it is not to be wondered at that the soluble products of the reaction differ in some respects, as will be explained in the second chapter.

It is now easy to understand why the previous coction of the fibrin hinders alike its solution in dilute hydrochloric acid and in carbolised water. It is because heat at 100° kills the microzymas as it destroys the activity of all zymases, and doubtless because the intermicrozymian albuminoid matter has undergone the special coagulation which hinders it from effecting the gelatinous combination with hydrochloric acid before spoken of.

Summary

Fibrin is not a proximate principle. It decomposes oxygenated water correlatively to a change in the zymas produced by its microzymas. This zymas is the agent of the liquefication of starch and of the changes undergone by its albuminoid matter, whether in dilute hydrochloric acid or in carbolised water, conditions in which its microzymas do not undergo vibronian evolution. In short, these microzymas, whether in the fibrin or isolated, are not the agents of the decomposition of the oxygenated water after the manner of ferments—that is to say, physiologically by a phenomenon of fermentation—but only as producers of the proximate principle which the oxygenated water changes as it changes hydrocyanic acid.

To complete a knowledge of fibrin and of its microzymas, I recall the facts that Estor and I, in our note, described an experiment from which we concluded that in the presence of pure calcic carbonate, and for so long as the microzymas of the fibrin continued to evolve, they behaved, as regards fibrin, as alcoholic ferment, and as acetic, lactic and butyric ferments.[14]

Among these experiments I will describe two, because they were conducted on a sufficiently large scale, the better to establish results.

The proportions of the materials employed were as follows: fecula of potatoes, 5 parts, transformed into starch in 85 per cent of water; pure calcic carbonate, 1 part; and fibrin, fresh, moist, newly prepared, 0.13 parts. The temperature of the oven was 35° to 40°C (95° to 104°F).

The two experiments were started on the 22nd of May. The next day, disengagement of gas commenced, a mixture of carbonic acid and of hydrogen. From the 8th day, the gas was analysed repeatedly, and was found to be composed as follows, in percentages:

Date	3/6	8/6	18/6	25/6	3/7	17/7	3/8	15/8
Carbonic acid	80	91	88	77	62	50	67	71
Hydrogen	20	9	12	23	38	50	33	29
	100	100	100	100	100	100	100	100

The gaseous mixture is thus seen to have varied with the complication of the reaction.

One of the experiments was stopped on the 10th of September for the purpose of making the analysis. There was still a large amount of fecula not transformed. The products of the fermentation were as follows:

Absolute alcohol	21cc
Propionic acid	12g
Butyric acid	150g
Crystallised acetate of soda	650g
Crystallised lactate of chalk	709g

The second operation, upon a greater scale, was continued until the lactate formed had been transformed; the analysis of the products was made on the 10th of May of the following year. The experiment then had lasted nearly a year. There was still some fecula not transformed. There were found:

Alcohol mixed with higher alcohols	78cc
Proprionic acid	80g
Butyric acid	680g
Acids higher than the butyric, up to caprylic	245g
Crystallized acetate of soda	725g

Thus, as in the classical lactic fermentations, the ferment which produced the lactic acid is also that which destroys this acid in the lactate of lime.

It is only necessary to observe that the products formed by the microzymas of the fibrin differ greatly, both in proportion and in quality, from those of ordinary lactic fermentations, and especially from those by mother of vinegar.

I shall, by and by, insist further on the fact that the bacteria of the microzymas which evolve in the first have gradually but completely disappeared in the second so that at the end there only remained a few forms closely allied to the microzymas. But I insist here on the fact that for the two experiments 200g of fresh fibrin were employed containing at the start at most 0.2g of microzymas to effect the prodigious transformations of the fecula. The fibrinous microzymas are therefore ferments of rare energy.

Such were the preliminaries to the discovery of the third anatomical element of the blood. For a complete understanding of the fibrin and the products of its changes, it is necessary to know in what light to regard the albuminose of Bouchardt, which this savant believed existed in the supposed solution of fibrin in dilute hydrochloric acid, and to do this we must have a better knowledge of the albuminoids.

CHAPTER ONE

NOTES

1. Thénard, *Traité elementaire de chemie*, Vol.1 p.528. 6th Ed., 1834.
2. See Memoir, *Essay sur les substances albuminoid*. These de la Faculte de Medecine de Strasbourg, 1856.
3. *Annels de Chimie et de Physique*, 3rd S. op. Vol. LIV, p.28 (1858).
4. In 1858, Pasteur believed so little in the existence of germs of ferments in the air that he asserted that the lactic ferment and beer yeast were born spontaneously of the albuminoid matter of the fermentable media.
5. [Evan Landois, in his *Physiology* (Eng. trans. by Stirling, 1889) makes this mistake. He makes no reference to its ferment—Trans.]
6. *C.R.*, Vol. LXXVI, p.654.
7. *C.R.*, Vol. LXIII, p.451 (1866).
8. [For explanation of the action of the microzymas of the calcareous rocks see *Role de la craie dans les fermentations*, Bull. Soc. Ch., Vol VI, p.484 (1866); also *Les Microzymas*, third conference; also post-chapter 8 of this work—Trans.]
9. Thénard, *Traieé élémentaire de chemie*, Vol. III, p.430 (1815).
10. *C.R.*, Vol. XIV, p.962 (1842).
11. It was this gelatinous mass that Thénard correctly regarded as a hydrochlorate of organic matter.
12. *C.R.*, Vol. XCV, p.926 (1887). I have since further studied this subject. Hydrocyanic acid and oxygenated water react upon each other; at first without the liberation of gas; an oxamide is formed which crystallizes; at the same time there is a liberation of heat, which increasing, the oxidation is accomplished with production of urea and liberation of oxygen. It is then solely because hydrocyanic acid and oxygenated water react first of all, that the microzymas of fibrin are protected, and not as has been supposed, because hydrocyanic acid acts as a poison upon these microzymas. The fact that after the destruction of the hydrocyanic acid the fibrin again decomposes oxygenated water proves that what happens is not a phenomenon of poisoning. This will be further treated hereafter.
13. [For the purpose of continuing the 'conspiracy of silence' beneath which the discoveries of Béchamp have been obscured for so many years, this word and its congeners are never used in the writings of the chief conspirators, nor in those of the numerous leaders of the profession who have been duped by them.—Trans.]
14. *C.R.*, Vol. LIX, p.717-716.

CHAPTER TWO

On the actual specific individuality
of the albuminoid proximate principles.

The albuminoids.

The phenomenon of coagulation.

The albuminoids of the fibrin.

The albuminoids of the serum.

Haemoglobin.

Haemoglobin and oxygenated water.

To solve the problem of the spontaneous coagulation of the blood, it necessary to know not only the three anatomical elements of this humour, but also the composition of the medium in which they live, because there are to be found the conditions of their existence.

Let us propose that, in accordance with the hypothesis of Hewson, Milne-Edwards and Dumas, fibrin does not exist dissolved in the blood, and further, that it is connected with what we have called fibrinous microzymas. We then recognize that the really liquid part of the blood contains all its components in a state of perfect solution, as in the serum separated from the clot.

In 1815, it was supposed that the serum of the blood contained albumin as the only albuminoid matter, and this was not only identified with the white of an egg, but with the albumen of the serous fluid of the pericardium and of the ventricles of the brain, with chyle, and even with pathological serous fluids, such as that of dropsy, of blisters, etc.[1] And these identifications were based solely upon a single characteristic: coagulation.

Even today, it is contended that two solutions contain the same albumin when they are coagulable at about the same temperature. But the phenomenon of coagulation has been so abused that it has become necessary to define it accurately.

CHAPTER TWO

The Phenomenon of Coagulation

At first, the term coagulation was applied to the transformation of the blood from a liquid to a solid state, in the same sense that one used the term to describe a liquid that solidified, or a vapour which condensed into a liquid. Fourcroy said of the white of egg, of the blood serum, etc, that they are concrescible[2] by the application of heat because they contain albumin. But over time, to the notion of coagulability, chemists added that of insolubility; to coagulate became for albuminoids the correlative to becoming insoluble. For instance, when the white of egg forms into a solid mass in a hard boiled egg, it is said to have coagulated, to have become at once solidified and insoluble throughout; but as will be seen presently, it is not so with the blood when that is said to be spontaneously coagulated.

When coagulation was thus strictly defined in a chemical sense, the insolubility of the coagulated substance was only considered relative to water as the solvent; solubility before coagulation was also relative to water. But we shall see that the idea should be completed by extending it to other solvents.

In the present state of science, for instance, the name fibrin is given not only to that which I have just studied, that of the blood, the general phlebotomy of adults, but also to that of the arterial or venous blood, without regard to the region of the vascular system from which it is taken, and without distinction as to age; also that of the chyle, that of the lymph or even of pathological serosities. And this fibrin was regarded as coagulated albumin without regard to the special action of fibrin upon oxygenated water, nor, as we shall, see, of its own coagulability.

A rapid review of the history of albuminoid matters will enable us to understand how, in 1875, it came to be supposed that fibrin was only a stage in the transformation or alterations of albumin.

Under the influence of Guy-Lussac and Thénard, and of Mulder and Dumas, chemists had defined a certain number of nitrogenous matters of animal or vegetable origin as specific, not only when they were a little different, but even apparently identical in their centesimal elementary composition. These matters Dumas called "neutral nitrogenized matters of the organization," recalling thereby an old classification of Thénard. Finally, they were called albuminoids, comparing them to albumen, or white of egg, taken for a type, because of certain properties and of some resemblances in composition.

The notion of specificity prevailed up to 1840; after that, in spite of Berzelius, the singular idea of the substantial unity of these substances seemed to prevail. This is how it came about.

It will be remembered that Bourchardt gave the name of albuminose to the fibrinous matter dissolved by very dilute hydrochloric acid. The reason for the invention of this new word is a curious one.

Biot had observed that the watery solution of the white of egg deviated the plane of polarization of polarized light to the left; Bourchardt, having found that the hydrochloric solution of fibrin also deviated the same plane of polarization to the left, concluded that "as the soluble principle of fibrin is identical with the dominant matter of the albumen of the egg, I propose for this pure substance the name of albiminose."

Then, dissolving in very dilute hydrochloric acid various other analogous substances thus obtained, he generalized as follows:

> "The fundamental principle found in the fibrin, in the albumen of egg, in the serum of blood, in the gluten of cereals, in casein, is always the same; it is albuminose, mixed or combined sometimes with earthy matters, phosphates of lime and of magnesia, sometimes with alkaline salts, sometimes with fatty matters, which mask their essential properties.
>
> If this ephemeral combination be destroyed by a really inappreciable proportion of acid, the albuminose solution is then found to have identical properties, exactly similar chemical reactions, similar action on polarized light, always deviating to the left, the energy whereof, other things equal, is always proportioned to the weight of the substance dissolved.[3]

The above amounts to saying that the albumen of the white of egg, that of serum, the essential matter of gluten, of casein and of fibrin, are the same substance, possessing the same rotatory power.

We shall see how, even as to fibrin, to what extent the

observation of Bouchardt was superficial and how he deceived himself in generalizing it. He deceived himself so strangely that he did not think for a moment that he had to do with hydrochloric combinations, believing that the quantity of hydrochloric acid of his solvent was inappreciable, etc.

The chemists were equally careless. Gerhardt adopted Bouchardt's point of view and extended it.[4] In Germany, especially, a legion of chemists maintained the substantial identity of these matters; P. Schutzenberger (a native of Holland, domiciled in France) adopted it. It was because they knew very little about the chemical constitution of albumen; so little that Gerhardt consigned albuminoid matters to a place below asphalts and bitumens, and in the general confusion Naquet thought that albuminoid substances did not belong to the domain of chemistry, but to that of physiology, as remains of organs.

But in 1856, while I was busied with the researches which resulted in the discovery of the microzymas, in a work on the source of urea in the organism,[5] by arguments drawn as much from chemistry as from physiology I had established the specific plurality of the albuminoids and demonstrated that these substances, animal and vegetable alike, produce urea by decomposition following a phenomenon of oxidation.

In this work I succeeded in expressing the chemical constitution of albumin and of the albuminoids in general, regarded as proximate principles. I showed that their molecules were very complex, the most complex known, inasmuch as formed of numerous non-complex molecules of the fatty and aromatic series, among which were amide derivatives, amides and sulphides, in the number whereof urea was never wanting, so that if the ureides of Grimaux had been known, I should have said that albumin is a very complex ureide. In this work I laid the foundation for the future researches which led me to the discovery that the albuminoid matters, even those regarded as proximate principles, are either mixtures, like the albumin of white of egg, or organized things, like fibrin and vitellin.

The researches whereby I demonstrated analytically that there are a great number of natural albumins and albuminoids, reducible to rigorously defined proximate principles, were made the subject of examination by a commission of the Academy of Sciences and of a report by Dumas.[6]

The memoir which is the subject of this report contains the demonstration of the specific plurality of albuminoid matters, and that the doctrine of their substantial unity is an error.[7] Among other things, I demonstrated that the classical albumin, the white of egg of the fowl, the type to which had been referred all those matters which were identified under the name of albumin, was a mixture of three proximate principles, irreducible to one another; all three albuminoids, all three soluble and deviating the plane of polarization of light to the left, whereof two are coagulable by heat, the third not coagulable, a veritable zymas.

And J. Béchamp, having analyzed by the same method the whites of eggs of a number of oviparous animals, birds and reptiles, discovered among them other albumens, other zymases, different from those of the egg of the fowl; so different and differing among themselves that he was able to specify the species of a bird by the albumens of its egg.[8]

But prejudice and partisanship are so tenacious that nothing was of any avail. Notwithstanding the report of Dumas, long afterwards, a learned physiologist held that fibrin was a proximate principle. He did so in reliance on the opinion of Duclaux, proclaiming "the extreme mutability of albuminoid matters and the folly of the chemical specifications established in this category of organic substances",[9] and again maintained that fibrin was a proximate principle.

It is upon such opinions that rests the assurance that fibrin is a stage in the mutations of albumin and that the albumen of milk is a consequence of another change in casein, as asserted by Declaux. All this is inaccurate and one may even say absolutely untrue, for pure albuminoid matters are fixed, and are as rigorously definable and specific as any other proximate principle.

Independently of the ignorance which prevailed regarding the chemical constitution of the albuminoids, that which most contributed to perpetuate these prejudices was that so little was known concerning the faculty of the albuminoids to form

CHAPTER TWO

combinations with bases or acids. Even Dumas held them to be neutral nitrogenous matters. It is true that Bouchardt said that they form combinations with the alkalines and alkaline earths, but he said that such combinations were only ephemeral. Thénard proposed the formation of combinations with hydrochloric and sulphuric acids, but no one paid any further attention. Lieberkuhn regarded the albumen of the white of egg as an albuminate of soda, but said also that casein was an albuminate of potash.

These kinds of combinations, under the hypothesis of substantial unity, served to explain the differences presented by these matters, compared with one another, as being soluble or insoluble. What is certain is that, at least in the animal organism, albuminoid matters are always combined with an alkali or an alkaline earth, and that further, these combinations are complicated by the presence of phosphatic earths, which they dissolve.

And as if to augment the confusion and force of prejudice, natural coagulations were proposed. At the same time that the insolubility of fibrin was sought to be explained by its combinations with phosphates, it was called coagulated albumen. As to the soluble albuminoids, to differentiate them coagulation by heat was employed; those which coagulated at the same temperature were regarded as identical. Casein was said to be insoluble by heat, but coagulable by acids, thus confusing a purely chemical phenomenon of precipitation with a physical phenomenon.

My researches have solidly established that from those natural materials which always constitute mixtures there can be separated by means of analysis the albuminoids, proximate principles (which when isolated have an acid reaction and which unite with bases in as definite proportions as any acid, so that casein produces with sodium a neutral caseinate), and a bicaseinate, which reddens litmus paper.

I have also demonstrated that these substances can form combinations with hydrochloric acid and with acetic acid in several proportions. From these various combinations the albuminoid matter, whether soluble or insoluble, can always be isolated with its own proper characteristics and always with the same rotary power.

But the natural albuminoid matters, even when reduced to

proximate principles isolated from bases and other mineral matters with which they had been combined or mixed, are neither crystallizable, volatile nor fusible; they possess then none of the so-called constant characteristics employed by chemists to ascertain at once their purity and identity.

How then can one make sure that the substance isolated by analysis is always identical with itself? I employed as a constant the rotary powers employed for a like purpose by Bouchardt with the substances studied by him.

The following table gives the rotatory powers of the chief albuminoid matters on which Bouchardt experimented and disposes of the theory of the substantial unity of these matters. In the table the numbers are relative to the perceptible tint according to Biot.

CHAPTER TWO

ALBUMINOID MATTER	SOLUTION	ROTATORY POWER
White egg of fowl, whole, purified	water solution	−43°
First albumen of white of egg	watery solution	−34°
Second albumen of white of egg	water solution	−53°
Leukozymas of white of egg	water solution	−79°
Albumin of ox blood serum	water solution	−61° to −63°
Hemazymas of ox blood serum	water solution	−57.7°
Casein	like solution	−117°
	hydrochloric acid	−108.6°
	ammoniacal solution	−118°
Lactalbumen of cow's milk	acetic solution	−66°
Galactozymas of cow's milk	water solution	−40.6°
Gluten of wheat, whole	dilute hydrochloric acid	−101.4°
A fibrin of gluten	acetic solution	−88°
A glutine	water solution	−109°

We will now see how it is with a solution of blood fibrin in very dilute hydrochloric acid.

The hydrochloric solution of fibrin, separated from its microzymas, contains a mixture of albuminoid matters, soluble and insoluble in water.

The limpid solution, which has been obtained with or without the addition of phenol, has a decidedly acid reaction and is without action on oxygenated water. The solution is really one of hydrochloric combinations with albuminoid matters, whereof the greater part is insoluble in water. In fact, on the addition of dilute ammonia so that the liquor becomes faintly alkaline, an abundant dead white flocculent precipitate is produced which, collected on a filter, well washed with distilled water, alcohol and ether, and then rapidly dried in a dry vacuum, forms a pulverulent matter. Was this the whole of the fibrin less its microzymas? If so, the fibrin is purely and simply dissolved; if not, the solution was the result of a reaction. The facts of the matter will be determined by dosing.

A manipulation of 60g of fresh fibrin, containing 11.5g of matter, dried at 100°C furnished 7.6g of this insoluble matter likewise dried at 100°C; that is to say, only 66% of the weight of dry fibrin. Consequently, we can say that 34% of the matter remained in solution. If the reaction is continued longer before separating the microzymas, the quantity of matter precipitated by the ammonia diminishes, while the dissolved portion increases.

The substance insoluble in water—the ammonia precipitates—possesses further the same elementary composition as fibrin, but it differs from the intermicrozymian substance in that it is directly soluble in very dilute hydrochloric acid, as well as in acetic acid and in ammonia. I have given it the name of fibrinine.

Further, the fibrinine does not decompose oxygenated water and does not liquefy fecula starch.

Among the substances which ammonia does not precipitate is one which alcohol precipitates after the separation of the fibrinine. This precipitate is a mixture; one portion is soluble in water, but the other does not redissolve. I have given the name fibrimine to that portion which is finally soluble in water.

The part precipitated by alcohol is the smaller part of the material which ammonia does not precipitate; the rest is to be likened, more or less, to the extractives such as are found in gastric digestion. I add that the fibrimine possesses the property of liquefying starch and I regret that I did not think of finding out whether it, or some of the compounds accompanying it, has the property of decomposing oxygenated water.

However that may be, the following are the rotatory powers of the hydrochloric solution of the fibrin as a whole, and that of the fibrinine and fibrimine:

CHAPTER TWO

ALBUMINOID MATTER	SOLUTION	ROTATORY POWER
Whole fibrin (from blood of sheep, cow and pig)	hydrochloric solution	$-72.5°$
Fibrinine	hydrochloric solution	$-67.4°$
Fibrimine	aqueous solution	$-80°$[10]

A comparison of these various and different rotatory powers, answering to other properties, not less different, of the bodies which possess them is sufficient to show that the identification made by Bouchardt which led him to believe that there was a substantial unity among albuminoids, had no basis in truth. Nevertheless it was upon this identification and on the results of the elementary analyses made upon mixtures and not on real proximate principles that was based the opinion which regarded fibrin as coagulated albumen, or as a stage in suppoitious changes in the albumen of the white of egg.

Although it has been ascertained that this albumen, coagulated or not, did not set free oxygenated water, this enormous difference was disregarded, as was the fact which followed the incomplete solution of fibrin in dilute hydrochloric acid, whence it was obvious that fibrin was not a proximate principle. Hence it is not surprising that for a long time muscle fibrin was confounded with that of blood, and that even today the fibrin obtained from blood is regarded as being the same whatever be the animal or part of the vascular system from which it comes, even also the fibrins of chyle, of the lymph and of pathological serosities.

Denis (of Commercy) had already established that certain fibrins of the veinous blood were dissolvable in a solution of saltpetre (nitrate of potash), while others, including fibrins of arterial blood, did not dissolve in it.

Estor and I demonstrated that the fibrin of the blood of very young kittens liquefied and disappeared in the starch which it had liquefied, while its microzymas evolved. On the other hand, I found that the fibrin of ox blood did not dissolve under the conditions specified by Bouchardt, and that it was necessary to employ hydrochloric acid at 3 in 1,000.

In another experiment, the fibrin of the blood of a young chicken, treated with hydrochloric acid at a strength of 2 per 1,000, did not even swell up even after remaining a long time in the oven, and at the end of several days the acid had dissolved very little of it, and the liquid hardly produced a precipitate with ammonia. Nevertheless, this fibrin decomposed oxygenated water before treatment; it decomposed it also after treatment when the acid had been eliminated by washing with water.

In short, to become convinced that fibrin is a much more variable anatomical substance than a definite chemical principle which is always the same, it suffices to recall the former observations of Marchal de Calvi, of Magendie, and of Claude Bernard, as well as those of Birot and of J. Bechamp.[11]

We must then erase the fibrins from the list of proximate principles to see in them only what they really are, that is, microzymian false membranes. The intermicrozymian matter of these fibrins is probably not the same in all.

However that may be, it is certain that the intermicrozymian matter of the fibrin common to ox or sheep blood is not coagulated albumin; that it is naturally insoluble, dissolving in very dilute hydrochloric acid only by a sort of auto-digestion, whereof the microzymas it contains furnish the zymas; and not only is it not a coagulated albumen, but it is itself coagulable by heat, becoming incapable of combining with dilute hydrochloric acid and of being thereafter dissolved in it.

In his report to the Academy of Sciences, Dumas did not fail to call the attention of savants to the fact that fibrin owes its property of decomposing oxygenated[12] water to that part of it which is insoluble in dilute hydrochloric acid.

Shortly afterwards, Bert and Regnard published a memoir upon the action of oxygenated water upon organic matters[13] which raised delicate historical questions of chemistry, physiology and facts, which I could not leave unanswered. This reply was the subject of several notes.[14]

CHAPTER TWO

In my reply to Bert and Regnard, I had chiefly addressed myself to the following assertion they had made; that:

" ... the blood, even defibrinated, acted with great intensity upon oxygenated water and that this action seemed to be entirely contained in the serum; and, further, that ossein very clearly decomposes oxygenated water."

I observed also that the authors did not distinguish between the expressions *organic matters* and *animal matters*, which was in conformity with the then state of science. But I knew what to believe regarding the fact that defibrinated blood decomposes oxygenated water, and I had ascertained the nature of the proximate principle which was its agent.

And first let us place beyond doubt the fact that it is not the serum which, in defibrinated blood, has the greatest share in this decomposition.

The fresh yellow (citron) serum which is first pressed out of the clot unquestionably sets free oxygen from the oxygenated water, which might be due to morsels of fibrin remaining in suspension. But the same serum, filtered several times upon a filter lined with sulphate of baryta, acts less and less on the oxygenated water without ever entirely ceasing to do so. This is very simply explained by the secretion in the serum of the substance which, in the fibrinous microzymas, effects the decomposition; but when the serum begins to be red coloured, the action upon oxygenated water is incomparably more energetic, the explanation whereof is as follows:

The defibrinated blood contains the red globules, and these contain the red colouring matter and their own (special) microzymas. Much has been written upon this red matter which has come to be called haemoglobin; and which was at first regarded as being a mixture of a colourless albuminoid matter called globulin with haematosin. Much has also been written upon haemoglobin, up to maintaining that it is not an albuminoid because it contains iron.

It was Dumas who first studied the colouring matter of the globules of the blood as an albuminoid proximate principle.

I have studied haemoglobin from the same point of view as other albuminoid matters. Confirming that it exists combined with potash in the globules, I have succeeded in combining it with the oxide of lead under the form of haemoglobinate. But the haemoglobinate of lead, decomposed by carbonic acid, furnishes soluble haemoglobin in the state of an absolute proximate principle.[15]

The solution of pure haemoglobin is coagulable by heat and by alcohol; in both cases the coagulum is absolutely insoluble in water. The solution is of deep red colour, the alcoholic coagulum is of a brick red. Haemoglobin, even coagulated by alcohol, decomposes in the presence of alcoholized ether under the influence of sulphuric acid, into haematosin and a colourless albuminoid matter.

That settled, and to be more precise, and apropos to the communication of Bert and Regnard, let us recall that Thénard proposed that the action of organic tissues upon oxygenated water was of the same order as that of platinum, etc. Nevertheless he did not fail to point out that while these metals decompose an "infinite quantity" of oxygenated water, it was not the same with organic tissues and fibrin, some decomposing it for a long time, others for a shorter period.

In the first category he placed the tissues of the lung, the liver, the spleen and fibrin newly extracted from the blood. In the second he placed the nails, the fibro-cartilage of the ribs, the tendons, the skin; these, he said, "soon entirely ceased to act," and, much surprised, he sought an explanation of these differences.

We will presently learn that the differences pointed out by that illustrious observer related to the different nature of the microzymas of the tissues; meanwhile I will only remark that the most active organic tissues belong to the vascular and respiratory systems. But we must not forget that Thénard assumed fibrin to be an isolated animal matter; that is to say, a proximate principle of animal origin.

Let us then compare the action of fibrin in this respect with that of haemoglobin, which is really an animal proximate principle.

To illustrate: let us take 30 grams of fresh moist fibrin and 6 grams of fresh moist fibrinous microzymas.

CHAPTER TWO

In 48 hours, the 30 grams of fibrin will have set free 1,600cc of oxygen from 180cc of water oxygenated to 10.5 volumes of oxygen, i.e. 53cc of oxygen per gram of fresh fibrin or 0.193 grams dried at 100°C.

In 48 hours, the 6 grams of fibrinous microzymas will have set free 1,000cc of oxygen from 160cc of water oxygenated to 10 volumes of oxygen, i.e. 166cc of oxygen per gram of moist microzymas or 0.139g, dried at 100°C.

Now to the haemoglobin.

In one experiment 10cc of a solution of this substance, pure, containing 0.338g of matter and 4cc of water oxygenated to 10.5 volumes of oxygen, have set free 30cc of gas in three quarters of an hour and 34cc in 24 hours.

Further, as soon as the disengagement of the gas began, the liquor became cloudy, flocculent matter appeared, and at the end the discolouration was complete. The phenomenon then is correlative to a change and an oxidation, for the oxygenated water being able to set free 42cc of oxygen had only set free 34cc of it. The oxygenated water is, further, almost completely decomposed. If one operates with sufficiently large quantities, heat is developed and carbonic acid mixed with oxygen is set free.

Other products of the discolouration by oxidation of the haemoglobin are numerous, and among them are albuminoid and other soluble products, and at the same time an insoluble body containing iron.

Haemoglobin then, a proximate principle, decomposes oxygenated water, becoming changed in so doing like fibrin and its microzymas; but at equal weights the haemoglobin produces less disengagement of oxygen than do fibrin or the microzymas.

That which distinguishes the mode of being of the haemoglobin is that, even coagulated by alcohol and then heated to 210°C (248°F), it becomes still more discoloured in decomposing oxygenated water, with disengagement of oxygen, while cooked fibrin becomes inactive.

But the haemoglobin is reducible to a colourless albuminoid matter and into haematosin; that is to say, into two new proximate principles.

Now the colourless albuminoid matter of the decomposition, freed from the sulphuric acid with which it had been combined, does not set free oxygen from oxygenated water. On the other hand, the insoluble haematosin and oxygenated water react strongly with disengagement of heat and of oxygen mixed with carbonic acid, while, absorbing a part of the oxygen, it is entirely transformed into soluble products. And, quite the opposite of what happens with fibrin and fibrinous microzymas, free sulphuric acid does not hamper the reaction.

It is evident from this that the haemoglobin owes to the ferruginous molecule of haematosin, which is one of the constituent molecules of its own molecule, the property of decomposing oxygenated water, destroying itself by oxidation. And it is thus that certain proximate principles of the fibrinous microzymas and the oxygenated water react, causing the decomposition of the latter with the disengagement of oxygen.

Here then are many undisputed proximate principles which act upon oxygenated water after the manner of the organic tissues of which Thénard spoke, and after the manner of fibrin, which is also an organic tissue.

It is useful to connect the facts relative to haemoglobin and to haematosin with the reciprocal reaction of the hydrocyanic acid and of oxygenated water, to show that they are not isolated facts. Further, Thénard himself observed that oxygenated water of a certain concentration reacted upon cane sugar with disengagement of oxygen and carbonic acid.

In defibrinated blood, it is then especially the haemoglobin of the blood globules which is the agent of the decomposition of oxygenated water. If the lemon coloured serum (always with little intensity) effects this decomposition, it is because it contains, besides its own albumen, some proximate principle, zymas or other, which is able to do so. In fact the albumen of the serum,[16] isolated and pure, is as little endowed with this property as is the white of egg and the colourless albuminoid of the decomposition of the haemoglobin. This colourless albuminoid of the decomposition of the haemoglobin, by its rotatory power and other properties, is absolutely distinct from the albumins and albuminoids of the table.

CHAPTER TWO

But the blood globules also contain microzymas which decompose oxygenated water. In studying them it is necessary to observe that the organic tissues which effect this decomposition owe this power especially to their anatomical elements or to some proximate principle secreted by them. In other words, the property of decomposing oxygenated water does not characterise organic tissues or bodies, as was believed by Thénard.

The study of these albuminoids in general, and especially of those of the blood, proves that the nitrogenous intermicrozymian matter of the fibrin is of a special nature, distinct from all other albuminoid matters, especially from the type of albumin which may be coagulated, and that it is itself coagulable by heat, becoming thus absolutely insoluble in very dilute hydrochloric acid.[17]

But the special study of the fibrin which has revealed the fibrinous microzymas has taught us nothing with regard to the condition of the fibrin in the blood during life, that is to say nothing regarding the relation of the intermicrozymian matter and the microzymas. This will be the subject of the next chapter.

NOTES

1. Thénard, *Traite de chemie*, Vol. III p.432 (1815).
2. [Obsolete; from the Latin *concrescere*, to grow together; hence to solidify.—Trans.]
3. *C.R.*, Vol. XIV, pp.966-967 (1842).
4. Gerhardt, *Traite de chimie organique*, Vol. IV, p.436 (1856).
5. *Essai sur les substances albuminoides et sur leur transformation en l'uree*, These de la Faculte de medecine de Strasbourg (2nd S.) No 376 (1856).
6. *C.R.*, Vol. XCIV. The members of the Commission were Milne-Edwards, Peligot, Fremy, Cahours, Dumas reporter.
7. *Memoir sur les matirees albuminoides, Recuiel des memoires des savants etangers*. Vol. XXVIII, No. 3. Imp. Nat.
8. J. Béchamp, *Nouvells recherches sur les albumines normales et pathologiques*, I.B. Bailliere et fils, Paris (1887).
9. Dastre, *C.R.*, Vol. XCVIII, p.959.
10. To complete these comparisons, in order to give a better understanding of the specific individuality of each albuminoid proximate principle, and to show still further the value of the new method of research, which for shortness I call the antiseptic method, I add the following:

    We know that fibrin, left to itself in carbonated water, changes while dissolving in great part without becoming fetid, leaving a residue of microzymas enveloped in an insoluble albuminoid atmosphere.

    In short, while spontaneously transforming, fibrin produces some dissolved materials and others insoluble. The whole of the dissolved portion, albuminoid and others not volatile, had a rotary power -29° to -30°, which proves that under these conditions the soluble products are different from those formed in the change on contact with very dilute hydrochloric acid. Among the dissolved products, which together have the above rotary power, were one zymas and several soluble albuminoids, coagulable by heat and having different rotary powers, in fact, from those of the hydrochloric solution.

    More than ten years after the report of Dumas and the publication of my memoir on the albuminoid matters, a learned physiologist, Dastre, reached the same conclusions on applying the antiseptic method to the study of fibrin. He also found, in effect, that crude fibrin, "in contact with antiseptic salt solutions (flouride and chloride of sodium) does not merely dissolve, but is transformed" into diverse substances called globulines, proteoses, propeptones, peptones, as it does under the influence of gastric juice. Dastre found also that the spontaneous transformation of fibrin resulted in the formation of soluble and insoluble products without taking the microzymas into account, and further he generalized by applying the same method to crude albuminoid substances without other distinction, and without specifying the nature of the products formed, for the words peptone, propeptone, proteose, and globulines are applied to a great number of very different things. To show this assertion is well

CHAPTER TWO

founded, here are the rotatory powers of the soluble products of digestion of some albuminoid matters digested by the gastric juice of the dog:

Fibrin (ox or pig)	−64° to −66°
Primalbumin of white of egg of fowl	−42° to −48°
Albumin of serum	−63.9°
Casein	−101° to −112°
Gluten	−122.7°
Glutine	-134° to −140.5°

11. See on these subjects *Les Microzymas*, pp.233-258 and J. Béchamp's *Nouvelles Recherches sur les albumines Normales et pathologiques*, p93.

12. *C.R.*, Vol XCIV, p.1276.

13. *ibid*, p.1383.

14. *ibid*, p.1601, etc.

15. *C.R.*, Vol. LXXVIII, p.850 (1874) and *Memoire sur les Maiteres albuminoides*, p.270.

16. The albumen of the serum! The rotatory power of this albumen has been given in the foregoing table to distinguish it from the albuminoid substances which Bouchardt confused under the name of albuminose. But its specification is of such importance for an exact knowledge of the blood that it would have deserved a chapter to itself; but thanks to what has preceded, this note will suffice.

First, let us remember that Denis of Commercy (1856) supposed that the plasmin of the plasma was decomposed, after the bleeding, into concrete fibrin and dissolved fibrin, afterwards called metalbumen. So that, according to this hypothesis, the serum expelled from the clot contains this metalbumen and its own albumin. Denis though he could verify this hypothesis by isolating from the serum its dissolved fibrin or metalbumen in the following manner. When crystals of a sulphate of magnesia are added to the serum, this salt is dissolved in it and a time comes when the serum is so saturated that no more will be dissolved and a precipitate is formed. It was the substance of this precipitate, insoluble in a saturated solution of sulphate of magnesia, but soluble in water, which was supposed to be dissolved fibrin. One was the more sure of it, because, under like conditions, the white of egg, common albumen, gives no precipitate to sulphate of magnesia. Such is the experiment which led to the admission of plasmine and its reduction which would give the metalbumen, which would dissolve in the serum with its own albumen, supposed to be identical with the albumen of white eggs. But all this is erroneous. The blood contains no plasmin and the serum does not contain two albumins whereof one is metalbumen. In fact, J. Béchamp, in his *Albumines Normales et Pathologiques*, p.31, has demonstrated that the precipitate determined in the serum by sulphate of magnesia is the same substance, endowed with the same rotatory power as the serum mentioned in the table. Further, he proved that certain albumins of the bird's egg are likewise precipitated by sulphate of magnesia, as is known to be the case with certain pathological serosities, but the precipitates thus obtained from these pathological albuminous liquids, also called metalbumens, possess different rotatry powers than those of the albumens of the white of egg of certain birds. Whence the conclusion that there is not a metalbumine or dissolved fibrine.

Futher, the specification does not rest only on the difference in rotatory powers, but on all the properties taken together.

17. To explain how fibrin in its spontaneous changes may give birth to a great number of products of decomposition, it is well to add the following to what I have said regarding the complexity of the albuminoid molecule. It is commonly said that albuminoid matter is a nitrogeneous quartenary. But I have shown that casein, absolutely free from mineral matter, contains phosphorous, and as the casein in the mammary gland results from the transformation of the albuminoid matters of the blood, it follows that these are also phosphoretted; casein also contains sulphur, which was known, but was supposed to be accidental. Then the haemoglobin contained iron. An albuminoid molecule may thus contain besides carbon, hydrogen, nitrogen and oxygen, phosphorus, iron and sulphur—seven elements instead of four.

I have observed that in the albuminoid matters of the vitellin microzymas the sulphur does not produce sulphuric acid precipitable by baryta when they are oxidised by the hypermanganate of potash; in this it resembles Taurine.

CHAPTER THREE

The state of the fibrin in the blood at the moment of venesection and of the molecular granulations.

The fibrin without microzymas.

The haematic microzymian molecular granulations.

The experiments which demonstrated that fibrin was a complete whole, formed of a special albuminoid matter belonging to it, and of microzymas and not a proximate principle, did not solve the problem raised by Dumas, i.e. it did not ascertain under what condition a substance thus composed existed in the blood. Neither did it solve the problem of whether such a substance pre-existed in the blood, or whether it was the result of a chemical transformation accomplished after the bleeding.

It was not at the first attempt that I solved the problem in the sense of the conclusion arrived at in this chapter; that the blood really contains the fibrin in the state of microzymian molecular granulations, where the microzyma and the special albuminoid matter are closely associated in a very special anatomical element. The solution of this problem can only be explained after understanding the observations which have been covered in the first two chapters and those which I have yet to describe.

Let us recall what, in 1869, was the state of opinion regarding first the pre-existence of the fibrin in the blood; and secondly, its production in the blood after it had issued from the vessels.

On the pre-existence of the fibrin there were two opinions: According to one, which was that of Hewson, Milne-Edwards and Dumas, fibrin existed in the blood in a condition of extreme division, in fine molecules, which after the bleeding became consolidated to constitute the ordinary fibrin. According to the other, which was also that of Hewson, and, in a certain manner, of Dumas, it existed in a state of solution or of quasi-solution. Bernard[1] allied himself to this view by proposing that the blood contained an albumino-fibrinous liquid which could only remain liquid in the body, taking on the form of fibrin after the bleeding.

These opinions, nevertheless, have had no part in the solution of the problem. They had been so thoroughly lost to sight that Estor and I had tacitly ranged ourselves among those who did not admit the pre-existence of the fibrin. Indeed, after having proved the presence of the microzymas of the blood in the fibrin by the fact of their vibrionian evolution in the substance itself, we said, in February, 1869:

> "That which is called the fibrin of the blood is a false membrane formed by the microzymas of the blood, associated with a substance which they secrete in an acid state from the albuminoid elements of this liquid." [2]

It was because we had sought for and discovered the molecular granulations in the blood before proving that there were microzymas in the fibrin that we compared them to the microzymas of the liver, finding them smaller and more transparent than the latter.[3]

Nevertheless, these microzymas in the blood had not been isolated by us. The consideration that, in the liver, the microzymas are especially included in the hepatic cellules led us to seek in like manner for the microzymas in the globules of the blood. It was on this occasion that we made an experiment suggested by that observation, which is as follows:

Dr. Combescure, who, alongside of me, had experimented on the blood from the point of view of its supposed coagulation in the vessel by excessive consumption of alcoholic drink, had discovered that the blood, received directly from the vessels into alcohol at 40% to 45%, far from coagulating, dissolved in it.[4]

The experiment being repeated under the same conditions, we found that the mixture of blood and alcohol remained liquid, appearing limpid, depositing neither globules nor fibrin, but that by degrees it made an abundant deposit which the microscope showed was almost exclusively formed of molecular granulations animated by the brownian movement.[5]

CHAPTER THREE

It was from this result that we finally reached the solution of the problem, but only long after I had resumed the study of the fibrin and its changes. As I had done in 1869, I at first regarded the molecular granulations of the deposit from the alcoholized blood as being the microzymas of the globules. But when I had isolated the fibrinous microzymas of an extreme minuteness and after the study of the molecular granulations of spontaneously changed fibrin, liquefied, without becoming fetid, I doubted.

The following are the consequences of this doubt:

The third anatomical element of the blood and the molecular granulations of the blood globules.

The conditions of the experiment for isolating the third anatomical element of the blood are as follows:

Take alcohol, rigorously rectified, free from acid and from alkali, and dilute it with distilled water, to bring it to from 35 to 40 per cent. Into two volumes of alcohol thus diluted, one volume of blood is made to flow directly and without interruption as it comes from the vessels. So much for the blood as a whole.

To examine blood already defibrinated, it must be passed through a fine linen cloth, to remove the fibrin which might be held in suspension, and it is then poured into twice its volume of like diluted alcohol.

The mixtures, dark red, are left to themselves in a cool place, and there is formed by degrees a clear red deposit which takes at least 24 hours to be completed. The deposit is much more abundant for the entire blood than for the defibrinated. The deposit is first washed by decantation in alcohol at 35 per cent, then on a filter with alcohol at 30 per cent, until it is perfectly white. Under the microscope the matter resolves itself into an infinite number of very fine molecular granulations. These granulations are mixed with remains of cellules, more abundant in a deposit furnished by defibrinated blood.

I made several determinations of these molecular granulations. The following were made upon sufficiently large volumes of sheep's blood, by bleeding from the jugular vein.

800cc of the whole blood gave 37.4g of humid granulations completely drained, containing 5.76g of dried granulations per litre of the whole blood.

2,675cc of the same blood, first defibrinated, gave 22.1g of humid granulations, drained and containing 4.87g of matter dried at 120°C; i.e. 1.82g per litre of defibrinated blood.

However, these qualities are far from being constant, even for the blood of the same animal. For example, one litre of sheep's blood, in another experiment, gave only 5.70g of granulations dried at 120°C, and another, by whipping only, gave 3.15g of fibrin per litre, dried at 120°C.

However it may be with regard to the molecular granulations of defibrinated blood, let us consider them as representing (as will be demonstrated) the molecular granulations and the envelopes of the destroyed globules. The difference of 7.07g—1.82g = 5.25g will represent the molecular granulations which the blood without the globules will have furnished.

In the *Memoire sur les matieres albuminoides*, I still considered the molecular granulations as being microzymas such as they existed in the blood, and I showed that, like the fibrinous microzymas, they liquefied fecula starch and decomposed oxygenated water.

In fact, 1cc of humid molecular granulations of the entire blood disengaged in 12 hours 26cc of oxygen from 2cc of water oxygenated to 15 volumes of oxygen; and 1cc of humid molecular granulations of defibrinated blood disengaged in 12 hours 23cc of oxygen from 2cc of the same oxygenated water.

These granulations possess then the properties of the fibrin and of the fibrinous microzymas. But are they really isolated microzymas? Are they not precisely the fibrin, such as it exists in the blood? That which led me to put this question was, first, the observation of molecular granulations of the fibrin spontaneously changed and, second, that the weight of these granulations is often greater than the weight of the fibrin which the same volume of blood was able to furnish.

I then treated granulations obtained from the blood, diluted in alcohol as I had treated those of the altered fibrin or the fibrin itself to isolate from it the microzymas.

CHAPTER THREE

Treatment of the haematic molecular granulations with very dilute hydrochloric acid.

The humid deposit of the isolated granulations is treated with very dilute hydrochloric acid. In contrast with what happens to fibrin, they disappear almost instantaneously at the ordinary temperature, even when the acid is diluted to one in a thousand. After some minutes, ten grams were dissolved in a cloudy liquid from which microzymas, as small as those of the fibrin from whipping, were slowly deposited.

The limpid hydrochloric solution, separated from the microzymas and saturated by carbonate of ammonia, furnished a precipitate of fibrinine, which when well washed does not decompose oxygenated water, while the microzymas, isolated and washed, decompose it.

The rotatory power of the materials dissolved by the hydrochloric acid was $-74°$. As to the rotatory power of fibrinine, it has been found in acetic solution to be $-68.9°$.

These values are the same as those of fibrin and of fibrinine of fibrin under the same conditions.

The very great difference between fibrin in the state of molecular granulations and fibrin obtained by whipping resides then essentially in the manner of reaction to very dilute hydrochloric acid, the solution of ordinary fibrin being a function of time and of temperature.

The blood of the ox and that of the rabbit behaved exactly as did that of the sheep. But the blood of the duck offered interesting peculiarities, as we shall perceive presently.

The only condition for the success of the experiment, indispensable in every case, is that the shed blood shall flow directly from the vessel into alcohol at 35 to 40 per cent, the volume of which should be twice that of the blood to be collected. The least interval, as for instance receiving the blood into a porcelain capsule or one of glass, from which to pour it into the diluted alcohol is sufficient to compromise the result.

Under circumstances similar to the last, the deposit seemed to be made more rapidly and instead of being pulverulent, it was flocculent. The deposit, collected as usual 24 hours afterwards, and washed in the same manner, had already acquired something of the condition of the ordinary fibrin, not being dissolved immediately in dilute acid of the usual strength, but requiring at least 48 hours for the cloudiness to subside under the same conditions under which (when poured directly into the alcohol) it was effected at once.

Hence it results that the matter of the molecular granulations of the deposit formed by the instantaneous mixture of the blood as it issues from the vessels with dilute alcohol is in a state chemically, physiologically, and anatomically nearest to that which it assumes in the circulating blood. It is, therefore, necessary to obtain the most exact idea of the physical constitution of the molecular granulations in that condition, in order to understand their constitution in the blood at the moment that this is held by its flowing into the alcohol.

It is an invariable fact that the molecular granulations of the deposit formed by the instantaneous mixture of the blood, at the time of the bleeding, with twice its volume of alcohol at 35-40 percent, are immediately dissolved by very dilute hydrochloric acid to a cloudy liquid containing the microzymas, and at low temperature.

It is also a fact that at the moment of the mixture with alcohol, the blood appears to be dissolved, so much so that a thin layer of the liquid is almost transparent; under the microscope no globules are seen, and it is with difficulty that one can perceive certain particles which are themselves translucent. The deposit, which is then made very slowly, is not the result of a formation first of all of a precipitate due to some reaction or coagulation of some dissolved matter; it is quite otherwise, since the globules which swim insoluble in the circulating blood are themselves dissolved, while being destroyed.

It is further evident that if the deposit had been produced by coagulation by the alcohol of a substance which existed dissolved in the blood, and which became insoluble in water from the fact of this coagulation, but became soluble in very dilute hydrochloric acid, the whole should be dissolved by the latter! But the microzymas are the persisting insoluble residue of the deposit as they are of the fibrin obtained by whipping.

But, further, this important fact must be borne in mind to which I have already called attention, that it is precisely when an interval, even a very short one, elapses between the venesection and the mixture of the blood with the alcohol, that the molecular granulations of the deposit are not immediately dissolved by the dilute hydrochloric acid; that is to say, that a certain coagulation has occurred.

CHAPTER THREE

We reach then the conclusion that the granulations of the deposit formed instantaneously in the mixture represent the nearest state to that which they have in the blood at the precise moment of the bleeding.

But what is the relation between the part of these molecular granulations which is soluble in dilute hydrochloric acid and the microzymas which remain undissolved?

It is the same which I have pointed out in the molecular granulations which exist in the deposit from the spontaneous alteration of the fibrin in carbolated water, *where each granulation is a spherical mass of albuminoid matter having a microzyma for its centre.*

In fact, the molecular granulations of the deposit formed by the alcoholized blood are like that; round, spherical, motile—that is to say, animated by the brownian movement, representing an exceedingly minute mass of albuminoid matter having a microzyma for its center. The very dilute hydrochloric acid dissolves the enveloping albuminoid matter, leaving the central microzyma undissolved.

A microzyma for nucleus, enveloped as by an atmosphere by a mass of albuminoid matter, insoluble in water, but which very dilute hydrochloric acid dissolves; such then is the physical constitution of a molecular granulation formed by blood diluted in two volumes of alcohol at 35-40 per cent. It may be called a microzymian molecular granulation.

But do molecular granulations thus constituted exist anatomically? And do any such exist in the blood? Yes, and the example is not solitary, but the haematic microzymian molecular granulation, with its special albuminoid atmosphere, is a prime example of this kind.[6] It only remains to represent that state of this atmosphere in the blood.

I revert to the remark, already made, that the deposit in the alcoholised blood is not the result of the precipitation of some dissolved substance contained in the blood, as per the plasma hypothesis. Direct observation had already permitted Estor and I to declare that the blood, as it issues from the vessels and before the commencement of the formation of the clot, contains around the globules an innumerable number of microzymas (that which we took for microzymas), most readily to be seen in the blood of very young animals—for example, of kittens from three to forty days old; and these microzymas we found

THE BLOOD AND ITS THIRD ELEMENT

to resemble those of the liver, but to be more transparent.

We did not fail to add that the reason they had escaped the attention of histologists was because of their minuteness and transparency.[7] It was in reality a dirigent idea which enabled us to find them where they had not before been seen. In the defibrinated blood they cannot be found.

That which we took for transparent microzymas, visible with difficulty, were the haematic microzymian molecular granulations, the same as those of the deposit in the alcoholised blood, except that the albuminoid atmosphere of the latter is a condensed atmosphere, contracted, become opaque, while in the blood it is inflated, soft and mucous, hyaline, and can again, as will be shown, become inflated in blood to which water has been added.

The theory of the phenomenon presented by the blood diluted in the alcohol is as follows:

When the blood is directly received into the alcohol under the specified conditions, its anatomical elements are rudely placed in new conditions of existence; while the globules are destroyed and their coloring matter (the haemoglobin) dissolves, the soft and mucoid atmosphere of the insoluble microzymian molecular granulations are deposited.

It is because the mucoid albuminoid atmosphere, insoluble in water, is condensed and hardened, laid hold of as it were, by the alcohol before being coagulated, that it is dissolved immediately in very dilute hydrochloric acid, while it is modified by coagulation and becomes insoluble as it does in the case of fibrin obtained by whipping, if some interval of time separates the venesection from the mixing of the blood with the alcohol. Once more, the condensed atmosphere of the microzymian molecular granulations contains the albuminoid matter in the state nearest to, if not identical with, that which it had in the blood.

The following experiment will instruct us still better as to the nature of the mucoid atmosphere of the haematic microzymian molecular granulations.

CHAPTER THREE

Experiment upon blood diluted in a saturated aqueous solution of sulphate of soda.

It is known that blood mixed with several times its volume of a saturated solution of sulphate of soda yields no clot, and that the globules are deposited in the mixture without yielding up their coloring matter.

Why under these conditions is no clot formed? The following is an attempted explanation of the phenomenon. The experiment ought to be made in winter in freezing temperature.

A volume of sheep's blood is received directly from the jugular vein into four times its volume of a saturated solution of sulphate of soda and the mixture left at rest. Twenty-four hours afterwards, the greater part of the globules will be deposited. The supernatant clear liquid is passed through a filter lined with sulphate of barium,[8] in order to retain the globules and the microzymas which remain in suspension. The filtration is necessarily slow. The filtered liquid, almost colorless, is absolutely clear. Mixed with oxygenated water, it slowly sets free a little oxygen.

The liquid which had remained limpid during the whole period of filtration (about 20 hours) furnishes by agitation a small mass of fibrin of a brilliant whiteness, having the membranous appearance of the fibrin obtained by whipping. The liquid separated from this matter, again filtered on a filter lined with sulphate of baryta, also sets oxygen free from oxygenated water.

The fibrin separated from the filtered liquid of the mixture of blood and sulphate of soda, that is to say, a fibrin without microzymas, does not set free oxygen from oxygenated water, but dissolves in it.

A mass of this fibrin without microzymas, about 1cc being placed in 8cc of oxygenated water having six volumes of oxygen, did not set oxygen free even after six days of contact (at least not more that would have been set free from it without the addition), but the fibrin had disappeared; it was dissolved. And the solution was albuminoid, for on treating with Millin's reagent a white precipitate appeared which became red on slightly heating it.

As to the clear liquid separated from this fibrin without microzymas, it gave, on acetic acid being carefully added, a slight albuminoid precipitate which has not been further examined. But the liquor separated from this precipitate contained a soluble albuminoid matter precipitated by alcohol, which, in acetic solution, had a rotatory power of $-86°$, very different from that of seralbumine.

From this experiment we may conclude that the microzymian molecular granulations are, like the globules, insoluble in blood, where the conditions of their anatomical integrity exist united; but the blood being diluted in the solution of sulphate of soda, although the globules remain insoluble, the albuminoid substance which forms the soft atmosphere of the microzymian granulations is dissolved in the new medium, at least in part, doubtless undergoing some transformation. In fact, while by agitation a part is separated in the condition of an insoluble mass[9] having the membranous appearance of fibrin, another part remains dissolved and can be separated from it, and has a rotatory power greater than that of seralbumin.

The change is also evident from the fact that the insoluble matter of fibrinous appearance is dissolved in oxygenated water without setting oxygen free. And if the limpid liquid of the filtration before and after the separation of this fibrinous substance sets free a little oxygen from the oxygenated water, it is because some of the matter which, in the microzymas, effects this disengagement is diffused in it. Thus the direct experiment demonstrates that that in the fibrin which decomposes the oxygenated water are the microzymas; the intermicrozymian mass of the enveloping albuminoid matter does not decompose it.

This constitutes a verification of the facts established in Chapter One; so much so that it is unnecessary to add that the defibrinated blood, treated with sulphate of soda, does not yield any fibrinous substance.

Such are the facts. But these microzymian granulations which Estor and I took for microzymas, had they not been already perceived? On this subject the following is the only information I have been able to collect:

> "There is to be found", says Frey,[10] "in human blood, besides globules, agglomerations of small pale granulations 0.001mm to 0.002mm in diameter," adding that "these granulations, which had been noticed before, showed themselves sometimes with

CHAPTER THREE

active movements of protoplasm, sometimes with a molecular movement (brownian movement)."[11]

Further, these molecular granulations had been observed in other humors and animal tissues. Many opinions had been expressed as to their role, but it was not known what that role was, nor whether they were organized.

As a result of the anatomical analysis of blood treated with alcohol, the existence of the molecular granulations of the blood is certain. It now remains to explain how such an observer as Muller did not see them and could have maintained, under microscopic examination, that, except the globules, the whole of the rest of the blood was in a state of perfect solution.

To understand this, it is sufficient to take into consideration the anatomical and physical constitution of the haematic microzymian molecular granulations in the blood, i.e. a microzyma, whose diameter at the most is 0.0005mm, and which is enveloped in an atmosphere of a soft substance, mucous and hyaline.

But in the blood, this mucous atmosphere when much inflated may have the same refractive power as the surrounding liquid. It is not then surprising, the central microzyma being so very small, that it escaped microscopic observation; in fact, it only becomes visible when the albuminoid matter of the enveloping atmosphere, outside of the vessels, begins to undergo the allotropic modifications which cause it to acquire the properties which it possesses in the fibrin.

To convince oneself that this interpretation is a true one, it is sufficient to consider crystallin, the transparency of which is perfect. Nevertheless, anatomically, crystallin is constituted by two layers of crystalline tubes. It also contains, like other anatomical elements, a crowd of microzymas; all this the microscope is incapable of showing directly because all the parts in the entire organ have the same refractive index. But as soon as, by grinding, this organization is destroyed, the conditions of existence of the anatomical elements of the organ are altered, and the microzymas and crystalline tubes become visible.

THE BLOOD AND ITS THIRD ELEMENT

Summary

The facts of this chapter establish definitely that the blood contains a third anatomical element, as constant and as necessary as the globules, consisting of a microzyma enveloped in an atmosphere of a special albuminoid substance, insoluble in the sanguineous medium. This anatomical element, previously unrecognized because of its anatomical constitution, its location and its properties, I have named *haematic microzymian molecular granulations*.

And now, when one considers that the weight of these microzymian molecular granulations (the deduction made of the molecular granulations which are furnished by the same volume of defibrinated blood) represents very nearly the weight of the fibrin obtained from the same volume of blood by whipping, it becomes evident that ordinary fibrin is nothing other than the microzymian granulations, heaped up and soldered together, whose albuminoid atmosphere has undergone, outside of the vessel, an allotropic modification by virtue of which, from having been directly soluble in very dilute hydrochloric acid, it has become soluble in it, as a function of time and of temperature.

We shall see how the anatomic constitution of the haematic microzymian granulations and the properties of their enveloping albuminoid atmosphere explain at the same time, mechanically, the phenomenon of the spontaneous coagulation[12] of the blood and the production of the fibrin by whipping.

Meantime, let us say that the foregoing demonstrations destroy the hypothesis of plasma and verify and complete the conceptions of Hewson, Milne-Edwards, and Dumas regarding the existence of the fibrin in the condition of fine granulations in the blood.

CHAPTER THREE

NOTES

1. Bernard, *Liquides de l'Organisme* Vol. I, p.152 (1859).
2. *C.R*, Vol. LXVIII, p.408.
3. *Ibid.*, Vol LXIV, p.713.
4. Dr. C. Combescure, *Theses sur les effects therapeutiques des ammoniacaux*, p.82. Theses of the Faculty of Medicine of Montpellier (1861).
5. *C.R*, Vol LXX, p.265 (1870).
6. Before I discovered the microzymian molecular granulations with an albuminoid atmosphere, I had already observed some of another kind. In isolating the microzymas of the pancreas, as Estor and I had isolated those of the liver, I found them enveloped with an atmosphere of complex matter. Treatment with alcoholised ether and with water dissolved the matters of the enveloping atmosphere and the naked pancreatic microzymas became visible with their special minuteness and color. The vitellin microzymas of the yolk of the eggs of birds are also enveloped with a complex matter. When the sort of tissue which constitutes the yolk of these eggs is steeped in a good deal of distilled water, the intergranular materials dissolve and the vitellin molecular granulations are deposited, spherical, sometimes mixed with vitellin globules. Washing with water removes everything which can be dissolved in it; then treatment with ether and with alcoholised ether dissolves the enveloping matter, a sort of alloy, an amalgam of fatty bodies and lecithin. Finally, washing with water and again with ether yields the vitellin microzymas of a perfectly white color. The vitellin microzymas and the pancreatic exist then well enveloped, in the condition of microzymian molecular granulations.
7. *C.R*, Vol LXIX, p.713.
8. (from p.112) The lining with sulphate of barium is made as follows. The filter should be without a fold; it must be filled with liquid, into which a suitably diluted solution of chloride of barium has been precipitated by a similar solution of sulphate of soda; the filtrate is again poured on the filter until that which passes is perfectly clear. It must be so arranged that the bed of sulphate baryta be at least half a millimetre in thickness. Finally the filter is to be washed with a solution of sulphate of soda.
9. There is nothing to be surprised at in this spontaneous passage from the dissolved state to an insoluble state. There is a parallel case in a modification of amylaceous matter, which from a condition of perfect solution, passes by degrees, in the liquor itself, to an insoluble state.
10. Frey, *Traite d'histologie et d'histo-chimie*, Fr. trans. from the German by P. Spielman, p.120 (1877). Eng. trans. by A. E.Barker, p.108. N.Y., Appleton. 1875. Note 1 p.sup.
11. Frey, *loc. cit.*, p.121.
12. [The blood does not really coagulate. See the postface—Trans.]

CHAPTER FOUR

The real structure of the red blood globule.

The microzymas of the blood globules.

The blood globules in general.

An exact knowledge of the physical constitution and the anatomical structure of the red globule is of great importance to the scope of this work. Is or is not the red globule of the blood a cellular anatomic element, constituted by an envelope with its content; or is it a kind of naked anatomical element, as was said of the milk globules? A question which must be resolved, in order to obtain a clear idea of the role of the red globule!

Prevost and Dumas[1] admitted that there was an envelope to the red globule and later, Helne[2], by exact observation, demonstrated the reality of the existence of this envelope. In 1856, Kuss taught us in his course on physiology at Strasbourg that

> "...the blood globules are not bladders, but compact organs, solid in all their parts, the least aqueous of all the organs of the body."

More than twenty years afterwards, Frey, interpreting the general opinion, said:

> "To sum up, the globule may be regarded as a mass of gelatinous substance saturated with water."

And he added:

> "In spite of the difficulties of observation and the uncertainty of such examinations, some authors have pronounced themselves, during these later years, in favor of the existence of a cellular membrane." [3]

Dumas, as we shall see, not only considered them as having a cellular constitution, but as individual living beings, saying that the deprivation of oxygen was fatal to them. Such was not the

CHAPTER FOUR

physiologists' point of view, and I cannot give a better proof of this than the following:

I had compared the scintillating corpuscle of the pebrine to a cellule already acknowledged to be alive, such as that of the globule of yeast. Pasteur asserted that this was an error and, referring to it, made the following statement:

> " My present opinion is that these corpuscles are neither animals nor plants ... from the point of view of a methodical classification they should be placed rather alongside of globules of pus or blood globules, or still better, of grains of starch rather than alongside of infusoria and moulds;" [4]

and later, he added that the corpuscle

> " is a production which is neither animal nor vegetable and incapable of reproduction, and it must be placed in the category of those bodies, regular in form, which physiology has for some years distinguished by the name of organites, such as globules of blood, globules of pus, etc." [5]

It is evident that that which is neither plant nor animal has neither organized structure nor life; it has no content in an envelope. But that was the state of science, and that which Pasteur believed regarding the blood globule and regarding pus was believed of every other anatomical element, for instance, of the spermatazoids, comprised in the "etc" of Pasteur, and which he compared to grains of starch, under the name of organites i.e., simulacra of organs.

If I insist particularly on this savant's mode of viewing these things it is because he has specially occupied himself with the blood and with what happens to its globules during its spontaneous changes.

It is no idle question to inquire whether the blood globule is naked or covered with an envelope separate from its mass. The property of a living body as to whose organization there is no dispute, whether it relates to an animal or to a globule of yeast, is that it is limited in its form by a continuous enveloping membrane (distinct from its interior medium), which takes the name of *tegument*. Inside the organism, the interior organs are individualized by their own tegument.

Among bodies which can only realize the conditions of their existence in water or in aqueous media, the insoluble tegument, endowed besides with special osmotic properties, protects the content, or the interior medium, against being dissolved or from other like change. The blood globule, like beer yeast, is individualized by its tegument; hence it is an organ, and not an organite.

That which has given rise to doubt as to the existence of a tegument around the red globule is that blood when steeped in water seems to be altogether dissolved in it. The globules disappear so completely that under the microscope not a vestige of them can be discovered. Those who, like Dumas, proposed an envelope thought that it was broken and that this explained why the weight of the isolated fibrin of the clot was greater than that obtained by whipping. But, as we shall presently see, that apparent solution of the globules is only the osmotic issuance of the interior part, soluble in water, across the envelope. The cellular tegument, which remains entire, is invisible under the microscope only because its refracting index is the same as that of the ambient liquid.

Demonstration that the red globule is a real cellule having microzymas for anatomical elements.

The demonstration of the unbroken tegument of the red globule of the blood, steeped in water, consists of establishing that its refraction is different from that of the liquid which results from this mixture.

To do this, the blood, defibrinated or not, is to be mixed with an equal volume of a solution of about 15 per cent of creosoted soluble fecula; at the end of 24 hours one can see that the globules resist the contact with the water much better, the envelope being clearly visible.

The experiment is most striking with the blood of the duck; in one of them, which had lasted three weeks, successive washings with a solution of soluble fecula and with water removed all the colouring matter, leaving for residue colourless globules, in which the nucleus could be seen rolling about in the water-logged

CHAPTER FOUR

globule; sometimes even the colourless tegument could be seen to wrap itself around the nucleus. The tegumentary vesicle, after these washings, has sometimes become so pale that is was only visible after an addition of an iodide paint which colours it yellow; but it is not coloured by the ammoniac solution of carmin, nor by that of the picrocarminate.

I have also similarly experimented on the blood of a fowl, a pigeon, a frog, a dog, an ox, and of a guinea-pig. In all cases the vesicle is seen entire, but among the elliptical globules having a nucleus, those of the fowl deserve more attention. And it is to be noticed that, in experiments which take up a long time, the nucleus of the blood globules of a bird ends by being resolved into fine molecular granulations which can be seen in the colourless envelope which remains whole. In the emptied vesicles of the circular globules, nuclei are never to be seen.[6]

After the publication of these experiments, Profs. J. Béchamp and Baltus described the process of tinting by which in all cases the unbroken cellular tegument may be distinguished, even in blood merely steeped in water.[7]

But that is not all; if there is no cellule without an envelope, neither is there one without microzymas. The blood globules are, in fact, no exception. On one hand, as I have before mentioned, in certain experiments the decoloured vesicles of the blood vessels of the duck, emptied of their colouring matter, contained their nucleus reduced to molecular granulations. On the other hand, I have stated that the deposit of molecular granulations formed in defibrinated blood, with two volumes of alcohol at 35-40 per cent added, came from the destroyed globules of this blood. It is necessary then, by direct experiment, to put this beyond doubt, the process of which has brought to light some particularly interesting observations.

To solve the problem of the existence of these molecular granulations in the globules of the blood, these latter were isolated from the defibrinated blood, separated from all trace of fibrin by filtration on fine cloth, and four times its volume of a saturated solution of sulphate of soda or of an analogous salt was added to it.

Molecular granulations of the globules of ox blood.

The globules, being received upon a filter with the usual precautions, are washed once more with sulphate of soda, and are then treated with alcohol of 35-40 per cent, which ought to dissolve the colouring matter. This was, in fact, done by washing in this alcohol until discolouration was as complete as possible, and finally with water. The filter retains the molecular granulations mixed with the remains of the circular envelopes, possessing the appearance and properties of those obtained from defibrinated blood treated directly with alcohol. The peculiarities already presented by the globules of the bloods of the duck and the fowl induced me to repeat these experiments upon these bloods.

Molecular granulations of the globules of fowl's blood.

The globules, separated from the defibrinated blood of the fowl by the saturated solution of sulphate of soda, are received upon a filter and treated in the same manner as was ox blood, with alcohol at 34-40 per cent, until deprived of their colour. While the washing was made with alcohol, the colourless residue upon the filter seemed pulverulent, as in the case of the ox's globules; but by continuing with washing in water, the pulverulent mass was transformed into a mucous mass.

Then, much surprised, I made this further experiment.

The defibrinated blood of another fowl was treated as usual by twice its volume of 35-40 per cent alcohol. The pulverulent deposit of molecular granulations was obtained in the ordinary way. The deposit when received upon a filter and washed with weak alcohol became white, remaining pulverulent, but as soon as water was added, the matter assumed the mucous condition of the granulations of isolated elliptical blood globules.

The mucous mass sets free the oxygen from oxygenated water; it dissolves only with difficulty in hydrochloric acid at 2/1000.

As a consequence of this, it is clear that the albuminoid atmosphere of the microzymian molecular granulations of the blood globules of the fowl, which assume the mucous state in water, is formed of a different substance from that of the atmospheres of the blood globules of the ox.

CHAPTER FOUR

We have already seen that the fibrin of fowl's blood, yielded by whipping, is scarcely attacked by hydrochloric acid, and the granulations of the globules of the same blood which have become mucous have the same property.

Molecular granulations of the globules of the duck's blood.

The red globules of the duck, isolated from the defibrinated blood by sulphate of soda, are also treated on the filter by alcohol at 35-40 per cent. The most prolonged washing with this alcohol does not at all discolour the molecular granulations; much more abundant than in the fowl's blood, they remain, until the end, coloured a brownish red. Is it otherwise with those of the deposit of the same defibrinated blood treated with alcohol?

The defibrinated blood of the duck treated at the moment of bleeding with two volumes of the same alcohol rapidly gives a deposit much more abundant than does the defibrinated blood of the sheep. The deposit, formed chiefly of the molecular granulations, is not decoloured by the washings with alcohol and with water, and remains of a brownish red. Treated with very diluted hydrochloric acid, it yields coloured solutions.

The bloods of the fowl and the duck deserve more complete study than, to my regret, I have been able to make. For the results prove that the globules, elliptical in form and with a nucleus, differ not only by their forms from the circular globules, but further that they differ among themselves.

These two facts, and those relating to the blood of the ox, prove that from this time forth we have to study not only the blood, but the bloods, perhaps as much in the properties of their anatomical elements as in that of their albuminoid components, and especially of their haemoglobins.

However that may be, it remains proved that the blood globules in general are not only constituted upon the model of the perfect cellule, but that that by a sojourn of the blood in a solution of fecula the refracting index of the tegument of the globules is modified without modifying its osmotic properties.

Further, it has been demonstrated that the blood globules have, as anatomical elements, molecular granulations constituted like the haematic microzymian molecular granulations, as well as the circular blood globules, and the elliptic, and that the molecular granulations of one blood globule may differ from those of another. Knowledge of the physiology of the blood will benefit greatly from the study of other special cases.[8]

CHAPTER FOUR

NOTES

1. *Ann: de Chimie*, Vol XVIII, p.280 (1821), and Vol. XXIII, p.53 and 90 (1823).
2. *Anatomie generale*, Vol. I, p.459. Jourdan's trans. into French.
3. Frey, *loc. cit.*, p.123.
4. *C.R*, Vol, LXI, p.511.
5. *ibid*, Vol. LXIII, p.134.
6. *C.R*, Vol. LXXXV, p.712.
7. *ibid*, p.761 (1877).
8. This remark gives importance to the following: "The whole of the small family of Chameleons presents the singular exception that their blood globules are elliptical. But nothing like it has been found in other animals of the same class and nevertheless the bloods have been examined of more than 200 species chosen among all the natural subdivisions of the group, including even the marsupials and the monotrema, which, in certain respects, seem to establish the passage between the normal mammifera and the oviparous vertebrates. No adult bird is known whose blood globules are not elliptical. It is the same with reptiles, batrachians and ordinary fishes. Among cartilaginous fishes, the lampreys, for example, the form of the globules is nearly circular." (Milne-Edwards, *Physiologie comparee*, Vol.I, p.48, etc, 1857 ed.).

   The form, the external characters, among all living beings, are allied to the entirety of their other properties. Why should it not be the same with their blood globules and their other anatomical elements?

   [Milne-Edwards is correctly quoted by Prof. Béchamp, as above, but the statement of Milne-Edwards is not consistent with those of other authorities. All regard the chameleon as a reptile, and all say that the blood globules of all adult reptiles are elliptical.—Trans.]

CHAPTER FIVE

The real nature of the blood at the moment of a general bleeding.

The living parts of the blood protoplasm.

The unchangeable character of mixtures of proximate principles.

The vitellin microzymas and the blood globules.

The vascular system.

The blood is a flowing tissue.

The blood really contains three kinds of anatomical elements: the red globules, the white globules and the microzymian molecular granulations. Anatomically, the blood is constituted by three sorts of figured elements and by a fourth element, a liquid. Is this the serum, this liquid which is its interglobular and intergranular substance?

The three sorts of anatomical elements are living, insomuch as they are organized and contain microzymas which I have proved to be living by their function as ferments and by their capacity to become vibrioniens by individual evolution, which was a novelty for physiology and even for chemists.

Nevertheless, so far as concerns the red globule, since 1846 the statement that it is alive was not a novelty. In fact, in a memoir,[1] which merits much attention but is never quoted, Dumas made an observation which must be regarded as of the first importance.

He stated that to isolate the red globules in their integrity, by mixing the blood with sulphate of soda, a current of air must be introduced. Without this, they will change, losing their coloring matter which itself also changes. And he said:

> " The globules of blood act as though they were really living beings, capable of resisting the solvent action of sulphate of soda so long as they are alive, but yielding to this action as soon as they have succumbed to the

CHAPTER FIVE

asphyxiation which affects them by the deprivation of air, and which manifests itself with singular rapidity, either by their change of color or by their rapid solution."

Dumas asserted clearly that the globules breathe, and that account must be taken of their membrane in explaining the phenomenon of respiration. He also stated that the breathing of an animal has the object of furnishing oxygen to the globules of its blood, and also to expel "the products into which they convert it."

He also remarked that in the discussions and calculations regarding respiration, the blood had always been regarded as a homogenous liquid, while it was only the serum which possesses this quality. He in no way disregarded the part taken by the serum in the phenomenon of arterialization, but he insisted on the preponderant part taken in it by the red globules.

To understand the blood, one must place oneself in the order of ideas of the memoir of Dumas, but broadened. That illustrious savant did not recognize in the blood, nor did any one else at that time, other anatomical elements than the globules, but there is another. He saw in the blood only three nitrogenous organic matters: albumen, fibrin and the globules, but there are others.

I will add that, in the serum, he made allowance for the share therein of the phosphates and other mineral matters.

At the moment of a general venesection, the blood has been regarded as being that which it is in the vessels while it circulates in them, but as being a mixture of the arterial and venous bloods; and we have seen that at this moment the blood is so thoroughly regarded as being alive that it was regarded as certain that coagulation was its death.

The blood being alive, it is necessary to recognize, in accordance with the doctrine of Bichat, that as in all the rest of the organism, the only things living in it are the anatomical elements. That is to say, that of the four parts which constitute it, the three kinds of anatomical elements are the only things living in it; the fourth, the serum, or that which will become the serum, the interglobular and intergranular substance, fulfilling with regard to them only one of the conditions of existence.

But as this conclusion conflicts with the prejudices of the schools, it is necessary to know what those prejudices are in order to combat them, for they are the negations of the doctrine of Bichat and precisely contrary to it.

In fact, while it is asserted that the globules of blood, as with all the anatomical elements, are only organites, and neither plants nor animals (as Pasteur said), that is to say, not living although organized, it was insisted that that which in the blood is still called plasma was living, a liquid whereof all the materials are said to be in a state of perfect solution, i.e. without any anatomic, figured structure.

But it is well to repeat that such was the state of science just as it was before Lavoisier and Bichat, when the philosophical naturalist Charles Bonnet, speaking of the organization, called it "the most excellent modification of matter." Even in France, a conception more or less analogous to it, that of protoplasm, was preferred to the striking conception of Bichat. But protoplasm or its synonym, blastema, was considered to be organized living matter without structure.

Here is one of the most precise descriptions of such matter:

> "A completely homogenous, amorphous matter without structure can be regarded as organized substance if it is constituted of numerous proximate principles, united molecule to molecule by special combination and reciprocal solution, and however simple may be this organization, it is sufficient to enable one to say that it is alive." [2]

Van Tieghem said:

> "Protoplasm is a mixture with water, of a greater or less number of different proximate principles, in the course of continual transformation."

Huxley said:

> "All protoplasm is similar to protein; all living matter is more or less similar to albumen."

Cauvet said:

> "Protoplasm is a nitrogenous liquid, more or less flowing, composed of a translucent joining substance and of fatty and albuminoid granulations."

CHAPTER FIVE

Even Claude Bernard said:
> "In its simplest condition, life, contrary to the idea of Aristotle, is independent of all special form; it resides in a substance defined by its composition and not by its shape; the protoplasm."

Pasteur said:
> "Living organisms are composed of natural substances such as life elaborates them; the proximate principles of living bodies which possess faculties of transformation which are destroyed by boiling." [3]

These quotations are sufficient. Protoplasm is regarded as a pure mixture of proximate principles, that is to say, of materials of a purely chemical order. Cauvet and others, Frey, for instance, have observed the granulations of the protoplasm, but they were thought to be pure proximate principles. This mixture was declared by some to be part of a process of continual transformation. It was seen by Pasteur as endowed with faculties of transformation, but without other proof of what is precisely the point in question, i.e. whether such a mixture can spontaneously change, can alter itself, or give birth to any living being whatever, be it a cellule or a microzymas. If protoplasm were that which it was thought to be, the conception of Bichat would be purely chimerical.

I have incontestably demonstrated, in contradiction to the theory of protoplasm and against Pasteur, that every mixture, artificial or natural, of real proximate principles with water is, by itself, in every way unalterable, and incapable of giving birth to anything living; in short, as NOT being in a state of continual tranformation and as not possessing any faculty of transformation capable of producing in it any spontaneous alteration.

And if, in such a mixture, boiling destroys the "faculties of transformation" of some zymas, this latter had not been produced spontaneously, it was the product of a living organism. In short, if the mixture contains some proximate principle which can be altered by oxygenation, by absorbing oxygen from the air, this principle is itself the former product of a living organism through the reaction of a zymas.[4]

I have given positive proof of all of this while studying the conditions of the spontaneous coagulation of milk, which was said to be a pure mixture of proximate principles. Cow's milk, creosoted by a suitable dose to destroy the influence of the germs of the air and completely protected from all contact with the air, first becomes sour and then coagulates. After which, vibrioniens appear in it. If by filtration, by the process which I have indicated in the case of the blood, both the globules and all the milk microzymas of the creosoted milk are absolutely removed, the limpid liquid which results, containing all the proximate principles of the milk, under the same conditions, does not become sour and consequently neither coagulates nor permits the appearance of the vibrioniens. "The faculties of transformation" then resided in the anatomical element of the milk which had been removed by filtration, and not in the rest of its substance, which may be called the physiological serum of milk.

The physiological serum of the milk, which has the same composition as blastema or of protoplasm, is then naturally unchangeable and consequently not living.

It is the same with the fourth portion of the blood, which we will call the physiological serum of the latter. And precisely as the anatomical elements of milk are the agents of its spontaneous alteration, because they are living, so too the anatomical elements of the blood are, on several accounts, the agents of its spontaneous alteration, as will be proved in the following chapter. But first must be determined the physiological role of this serum, in which are realized the conditions of existence of the anatomical elements, globules and granulations of the blood, while it circulates and after it has been shed.

I understand the "conditions of existence" of an anatomical element (following Bichat's conception) to mean the preservation of its physical being at the same time with the integrity of its tegument and that of its content, preserved with its composition unchanged, which can only occur by the element finding in the medium in which it lives all the materials for its nutrition.

Take, for example, the red globules; we know that in blood, steeped in a certain quantity of water, the soluble contents of its globules are diffused by osmose, the teguments remaining whole;

CHAPTER FIVE

on the other hand, we know that in the same blood, steeped in several times its volume of a saturated solution of sulphate of soda, its globules remain entire, both tegument and content.

We can even steep the blood in its own serum, without the globules being altered, without any trace of the colored content being dissolved. And it is the same with the molecular granulations as with the globules, so that if in blood, steeped in the solution of sulphate of soda, a small part of their albuminoid atmosphere is temporarily soluble, as we have seen, it is absolutely insoluble in the serum and each granulation remains whole and independent, the same as each globule, and this constitutes one of the conditions of the circulation.

But to understand the circulation and the reciprocal influence of the vessels and of the elements of their content, a slight diversion into embryology is indispensable.

In studying the development of the fowl to ascertain the role of the microzymas of the vitellus in the formation of the anatomical elements and of the organs, Estor and I have shown[5] that the container and the content of the vascular system are born and developed simultaneously with the aid of the microzymas and the unorganized materials of the vitellus.

We have never seen globules in the body of the embryo before the establishment of the circulation; they are formed on the spot.

Thus the anatomical elements of the tissues of the vessels and the anatomical elements of the blood contained therein are born at the same time, by the microzymas of the vitellus as builders, in the unorganized intermicrozymian medium of the vitellus.

Hence it results that the serum of the embryonal blood comes into existence concurrently with the globules and the granulations, having the non-organized parts of the vitellus for their source.

To sum up, container and content are born at the same time, develop at the same time, and at the same time become what they are destined to be in the future.

The blood ought to be studied not only by itself, but as being to the vessels that which the content of a cellule or of an organ is to its tegument. The tegument of the vascular system consists of the various tissues of the arteries, the veins and the capillaries. It must also be borne in mind that the system is directly related to the heart, the lungs, the liver, etc, and that the lymphatics (the chyle vessels) communicate directly with it. And as the content of a cellule, of an organ, does not exist without the container, so also the blood does not exist without the vessels which contain it and which make of the whole system an organ in more or less direct relation with every part of the organism.[6]

It must be observed that if there is any difference between the anatomical constitution of the container of the various regions of the vascular system, there is also a difference in their content. Independently of the color there is more oxygen and less carbonic acid in the arterial blood than in the venous. In several regions differences have been obseved in the proportion of the number of blood globules to that of the leukocytes. Lehmann observed that if the blood obtained from the portal vein gives fibrin by whipping, that of the suprahepatic vein does not furnish any by this means, proving, as we shall see, that the microzymian molecular granulations of the two bloods differ in something, and Denis has already pointed out that the fibrin of the arterial blood is not identical with that of the venous blood.

Consequently it is physiologically evident that the anatomical elements, conceived as being personally and individually living from whatever part of an organism they may be taken, exist there only because the conditions of their existence are found naturally realized there. It is not otherwise with the blood; the conditions of existence of its anatomical elements are only realized, in each point of the circuit, while it is contained in the vessel and circulating.

It is ordinarily said that the anatomical elements swim in the lymph, the liquor sanguinis or the plasma. Those who with Milne-Edwards proposed the existence of finely divided fibrin said that it too floated in the serum.

Anatomically, may we continue so to regard the reciprocal relations of the three anatomical elements and of the fourth portion of the blood? And is it correct to say that at each point of the blood current there are molecular granulations and globules almost in contact with one another? Is it not more correct to say that the fourth part, the serum, is only the intercellular and intergranular substance of these anatomical elements which hinders their immediate contact, a situation analogous to that which is correctly said to exist between the anatomical elements of the other tissues?

But, if this relation really exists for the blood contained in the vessels, must we not then say that the blood not only is not a liquid, but that it is a tissue like that of the content of the spleen, the liver, or the kidney, which are more or less flaccid? The softness of the tissue of the content of the vessels is much greater, that is all; we must then say that the blood is a flowing tissue.

The flowing state of the blood tissue is related to:

1) the soft consistence (gelatinous, it has been called) and the elasticity of the globules, whose tegument is incessantly lubricated by the intercellular liquor;

2) the much softer consistence of the swollen albuminoid atmosphere of the microzymian molecular granulations whose density is nearly equal to that of the serum; and

3) the absolute insolubility of the globules and the molecular granulations in the intercellular liquor, which again contributes to their individual independence.

This general insolubility of the anatomical elements is assured, at every point of the circuit, by the stability and even the origin of the composition of the very complex intercellular liquor, resulting from the nutritive functioning of the anatomical elements of the container and the content, and at the same time by the matters contributed by the diverse organs with which the circulatory system is in relation, and especially with the respiratory apparatus.

At the moment that the blood is shed it may be regarded as being the same flowing tissue that it was in the vessels. Nevertheless, there is already a profound difference; it is not only a mixture of venous and arterial blood, but of the bloods of all the regions, whose anatomical elements are violently placed in new conditions of existence, very different from their physiological conditions.

We shall now see how this change in the conditions of existence rapidly determines the manifestation of the phenomena of coagulation and then of other alterations of the blood.

NOTES

1. Dumas, *Recherches sur le sang.* C.R. Vol. XXII p.900 (1846).
2. *Dict. de Méd.* Littré et Robin, art. *Organique* 1878.
3. C.R. Vol LXXIII p.302. See letter of Pasteur to Donné. Pasteur's manner of thinking was still that of Chevreul at the time (1810) of the foundation of his chair at the Museum; Chevreul said, speaking of living bodies, that they are organic bodies in contradistinction to inorganic bodies, which we term minerals. Buffon called minerals gross matter, admitting that there was a universally diffused organic matter which he termed organic molecules, but Buffon wrote before the time of Lavoisier. Chevreul spoke of the proximate principles of organic bodies which are the products of life. Pasteur, speaking of the same proximate principles, says that they are natural substances elaborated by life, which have powers of transformation, etc. It may thus be truly said that there was no idea of life as bound to a determined, structural form of living anatomical elements, according to the conception of Bichat. It is thus to be understood how Pasteur could class in the same category, as organites, the red globules of the blood and grains of starch. It is true that the amylaceous granule had been regarded as being a vesicle, but Biot and Payen had shown that it was solid throughout its mass, and I have proved, in my researches upon fecula, that it had neither tegument, nor microzymas, being wholly formed of amylaceous matter contaminated with a trace of albuminoid matter.

    In the microzymian theory, it is not life which produces or elaborates the proximate principles, but the anatomical elements are constituted into living apparatus by the microzymas, according to the same mechanism by which the fibrinous microzymas cause starch to ferment, and elaborate the numerous proximate principles which I have described as produced in that fermentation.

4. The zymases are never the products of the spontaneous alteration of an albuminoid matter, but are always the products of the physiological function of a living organism and of an anatomical element in the latter. See the article *Zymas*, in *Dictionnaire de la langue francaise*, Littré (1869).

5. C.R. LXXV, p.962 (1872). We were led to undertake this embryological experiment as the consequence of the following experiment of which Estor was a witness: The Mother of vinegar formed of microzymas, united among themselves by a hyaline intermicrozymian substance, is a membrane of mucous consistence with which we have compared the false membrane called fibrin; but it is so much vegetable that it is hardly nitrogenised. But in the "mother of vinegar," under the conditions in which one forces its microzymas to live, these become by individual evolution bacteria, or by association, manufacturers of cellules. It is the same with the microzymas of beer yeast, which, in certain media, act as lactic and butyric ferments, undergoing vibrionian evolution; while in others they reproduce the cellule of yeast and the normal alcoholic fermentation.★

    The microzymas, then, can be manufacturers of cellules by grouping themselves together, and being grouped, becoming enveloped with a tegument when the conditions of existence of these cellules are united. And it is precisely this which the vetellin microzymas do during embryonic development.

    This new theory of the origin of the cellule does not weaken the axiom of Virchow: *omnis cellulae cellula*. One cellule may be derived from another cell according to another mode, that is all.

CHAPTER FIVE

Consequently, when Pasteur said that the globule of the blood is an organite incapable of reproduction because it could not be cultivated like beer yeast, he was mistaken, not knowing any other mode of reproduction.

* For the developments of the theory of the microzymas, manufacturers of cellules, see the following publications: *Conclusions Concerning the Nature of Mother of Vinegar and of Microzymas in General*, C.R. vol LXVIII p.877 (1869); *Researches on the Nature and Origin of Ferments. Ann. de chemie et de physique*, 4th series Vol. XXIII p.443. And for the theory in its entirety: *Les Microzymas Builders of Cellules*, see: *Les Microzymas*, etc., M. Chamalet, 60 passage Choiscul, Paris, pp.431—463 and p.948.

6. [This original conception throws a new light upon the purpose and relations of the circulatory system, which I hope to enlarge upon in a future memoir.— Trans]

CHAPTER SIX

The real chemical, anatomical and physiological meaning of the coagulation of the shed blood.

Coagulation of the blood.

The blood of the horse.

The serum of the blood.

Coagulation of blood diluted with water.

Second phase of the spontaneous alteration of the blood in calcined air.

Experiment proving that oxygen has no share in the destruction of the globules in the defibrinated blood.

Spontaneous alteration of flesh.

Spontaneous alteration of milk.

Coagulation of milk.

Fermentation of the egg.

Spontaneous destruction of the cellule of yeast.

Spontaneous destruction of tissues.

Spontaneous alteration of the blood.

The blood is a flowing tissue; Bordeu had already remarked that it was flowing flesh. This, chemically, histologically and physiologically, is far from being true;[1] the only thing certain is that the blood, like the flesh, is a tissue and that both of them are spontaneously alterable, as are all tissues, when the natural conditions of existence of their anatomical elements are no longer realized.

For instance, in the case of the muscular tissue, cadaveric rigidity follows death very quickly, and, in the case of the blood, the formation of the clot follows closely upon its issue from the vessels.

It is not disputed that the phenomenon of the coagulation of the blood is spontaneous. The standard facts concerning this phenomenon are as follows: the defibrinated blood obtained by whipping does not coagulate spontaneously, and the globules remain intact in the liquor which has lost its peculiar viscosity.

The blood of oxen and sheep (I leave for the present the blood of the horse), received into a glass or metal vessel, seems to coagulate throughout its mass, uniformly from the periphery to the center, forming a single solid clot which follows the shape of the vessel into which it has been received. This clot contracts by degrees, up to a certain limit, expelling from it in so doing the serum of a lemon color, which thereafter becomes red-colored, getting gradually deeper, so that the contracted clot (withdrawn from the edges) floats in the serum which has been expelled from its primitive mass. As Haller has already said, the clot is formed by the network of fibres of the fibrin which imprisons the globules in its meshes.

It remains to explain these phenomena invoking only the chemical, physiological and anatomical facts studied in the preceding chapters. The necessary condition for the tissue to remain flowing is that the properties of the anatomical elements and their independence remain unchanged, and that their relations with the intercellular liquor remain constant, not only in the vessel, but also after the venesection.

We know the distribution of the globules in the blood, and how they pass, one by one, into certain capillaries.

The distribution of the microzymian molecular granulations is such that if the globules should disappear, they will occupy all the space which the globules occupied; that is to say, that the microzymian granulations exist in such a manner in the blood that the globules move in it, displacing them unceasingly. In short, they realize the conception of Dumas, when he said of the fibrin that it exists in a flowing condition in the blood—except that this flowing condition is molecular, attached, as we have seen, to each molecular granulation with a microzyma for its nucleus, forming a limited atmosphere around each. This albuminoid atmosphere is absolutely insoluble in the blood serum.

To understand that the number of microzymas of the blood is sufficiently large so that, surrounded by the atmosphere which constitutes them microzymian molecular granulations, they may occupy every point of the blood mass, even that of the globules which were driven away, it is sufficient to know that they exist there in innumerable quantity. This is proved in the following manner:

The fibrinous microzymas, that is to say, the blood microzymas, are, along with the pancreatic microzymas, the smallest I have observed. They assume, in their extreme minuteness, spherical form. The diameter of these microzymas probably does not attain 0.005 mm in the humid state.

This enables us to calculate that in 1 cubic mm, there are at least 15,250 million. A litre of sheep's blood furnishes 5.25g of dried molecular granulations, which nearly represent the weight of the fibrin that the same blood furnishes by whipping.

But the fibrin, supposed dried, contains 1/193 of its weight of dried microzymas; then 5.25g of molecular granulations, likewise dried, contains 5.25/193 = 0.0272g; that is to say, 27mg of dried microzymas per litre of blood, which represents a very much greater weight of humid physiological microzymas.

In taking this figure for the weight of the microzymas in the physiological condition of humidity, and 15 milliards per milligramme or cubic millemetre, it is seen that one litre of blood contains more than 27 times 15,000 million microzymas.

But their weight is in reality much less than this, because, humid, these microzymas can retain 80% of water; in the blood, enveloped with an albuminoid atmosphere saturated with the intercellular liquor, they certainly retain less, but in a manner to render legitimate the approximate calculations above given.

It will be interesting to learn the thickness of the albuminoid atmosphere which surrounds each microzyma to constitute the microzymian molecular granulations, such as it exists in the blood at the moment of venesection.

An approximate idea of this can be obtained by considering that

CHAPTER SIX

the volume of the spherical molecular granulations with a condensed atmosphere of the deposit formed in the blood which has had added to it twice its volume of alcohol at 35-40 degrees is about 50 cubic centimetres per 1000cc of blood. Making allowance for the space occupied by the globules, we may consider that the volume of the molecular granulations, before the condensation of their atmosphere, was about twenty times greater to occupy the entire space of the 1000cc of blood. It will presently be proven that they do in reality occupy it all.

The albuminoid atmosphere being thus swollen and saturated with the intercellular liquor, it can be understood that the great number of milliards of these molecular granulations are sufficient to occupy the entire space presented by the blood, provided that their density be very little greater, if not equal, to that of the intercellular liquor which isolates them from one another.

This state of the microzymian molecular granulations explains the sort of viscosity which belongs to the blood, and how the globules, whose density is greater, move about in it without being deposited and are only deposited very slowly in the ox's or sheep's blood when at rest; and we shall see how the exception presented by the blood of the horse confirms these considerations.

We have now to inquire whether, after the shedding of the blood, the conditions which I have mentioned as necessary for the blood tissue to remain flowing can still be realized.

And first it is evident that this tissue, bearing in mind that we are considering a mixture outside of the vessels, is no longer in its natural physiological situation.

In this new situation the intercellular liquor, in which are united all the soluble organic and mineral products of the denutrition of the anatomical elements, the containers, and the contents, immediately changes its composition. The disassimilated products, which have become nonusable, are no longer eliminated, and the usable can no longer be utilized or renewed; further, the anatomical elements of the flowing tissue, which have imperative need of oxygen to function properly, are more and more deprived of it. After having consumed all that was held in reserve in the flowing tissue and which the uneliminated products thus accumulating in it had been able to absorb, the oxygen thus consumed is not renewed by respiration. The first change then which happens in the shed blood is that which the intercellular liquor necessarily undergoes in its composition.

The microzymian molecular granulations are the first anatomical elements to be affected by this change of medium and of conditions of existence, and, as we have seen, this impression is so intense and at the same time so rapid that it manifests itself in a few seconds in the profound change which occurs in the albuminoid substance of their atmosphere. From being immediately soluble in very dilute hydrochloric acid, it becomes insoluble in it, dissolving in it only as a function of time and temperature while being transformed. It follows that this influence has the effect of coagulating this substance relatively with very dilute hydrochloric acid.

That settled, the mechanism of the formation of the clot is as follows:

The microzymian molecular granulations exist throughout the whole of the space occupied by the flowing tissue, except that which is occupied by the globules and the intercellular and intergranular liquor. Thanks to their density, though very little greater than that of the intergranular liquor, they approach one another and come into contact when at rest; their albuminoid atmospheres, soft and mucous, mingle together, while at the same time their substance undergoes the coagulation of which I spoke. These changes are so rapid that the globules, although much superior in density, do not have time to be precipitated, and are caught in the meshes of the network formed by the soldering of the albuminoid atmosphere which constitutes the fibrin and membranes, as already said by Haller.

Both the molecular granulations and the globules are so closely connected by capillarity to the intercellular liquor that at the time when, or some minutes afterwards, the clot is completely formed, or, as it is said, the coagulation is complete, the vessel containing it can be turned upside down without any of the liquid escaping. This, in fact, is what is to be expected from what I have said about the distribution of the molecular granulations in the flowing tissue

CHAPTER SIX

and that of the intercellular liquor around the three anatomical elements.

It is true that the objection might be raised, with some appearance of reason, that that is precisely what happens in the plasmatic hypothesis. But that hypothesis has never been verified. On the contrary, I have directly proved that plasma does not exist in the blood, but that the existence of molecular granulations with their central microzymas was certain, as was also that of the microzymas in the fibrin obtained by whippng. But the following are two phenomena which the hypothesis of the plasma cannot explain.

After coagulation, the clot by degrees spontaneously divides itself into two parts. That which, in the clot, encloses the globules, i.e. the network of fibrin formed by the soldering of the microzymian molecular granulations, contracts more and more, up to a certain limit, preserving the shape of the vessel in which the clot is moulded, and while the retraction takes place a part of the intercellular liquor is expelled, constituting what is called the serum, in which the retracted tissue is now immersed.

And the first serum thus expelled is transparent and lemon colored, but by degrees the oxygen which is dissolved in the intercellular liquor, as well as that which the globules contain, is consumed. Then is manifested the phenomenon observed and explained by Dumas in globules deprived of oxygen; they change, and their changed coloring matter is diffused in the circumambient serum which becomes more and more of a deep red color. This is what the plasmatic hypothesis cannot explain, if one regards the plasma as a liquid in which all the components are in perfect solution. Let us now give a more direct demonstration of this fact.

All other things being equal, the rapidity of the coagulation of the blood may vary notably from one species to another. The blood of the horse, under the same conditions as that of the ox or the sheep, is well fitted to verify the truth of the role ascribed to the third anatomical element of the flowing tissue.

It is known that the (shed) blood of the horse is divided by rest into two layers: the lower, called *cruor*, is formed by the globules; the upper, called *liquor*, contains the microzymian molecular granulations.

The upper layer, transparent but not limpid, flowing, posessing the peculiar viscosity, can even be decanted after the globules are precipitated and very quickly forms a clot in all its mass so that the containing vessel may be turned upside down without one drop of the liquor escaping; after which the retraction, with loss of transparency, is produced and the serum commences to be progressively expressed from it, as in the case of the bloods whose globules do not separate. Now this retraction would not be produced if dealing with a substance really dissolved which, in coagulating, should become insoluble in the same medium.[2]

The peculiarity presented by the blood of the horse may be due to the greater difference between the density of its globules and that of the intercellular liquor, and at the same time to a greater softness of the albuminoid atmosphere of the microzymian molecular granulations, which will be more swollen, and consequently, their mass surrounded by the serum more readily traversed by the globules. I have, therefore, compared, as being the only thing accessible for experimentation, the serums of blood of the ox and the sheep with that of the blood of the horse, with regard to their general composition.

The rotatory power of the organic materials, taken altogether, of the serum of the blood of the ox is $-52°.8$.

In 100cc of this serum, there were:

Amount of fixed organic matters (g)	8.72
Amount of fixed mineral matters, ashes (g)	0.68
Sum of fixed matters (g)	9.40

Proportions in hundredths of organic and mineral matters:

Amount of fixed organic matters.	92.76
Amount of fixed mineral matters	7.24
	100.00

CHAPTER SIX

The rotatory power of the organic matters, taken altogether, of the blood of the horse is −53°.

In 100cc of the serum of the blood of the horse, there were:

Amount of fixed organic matters (g)	6.70
Amount of fixed mineral matters, ashes (g)	0.06
Sum of fixed matters (g)	6.76

Proportions in hundredths of organic and mineral matters:

Amount of fixed organic matters	99.1
Amount of fixed mineral matters	0.9
Sum of fixed matters	100.0

The rotatory power of the serum of sheep's blood was −64°.

The proportions in hundredths of the fixed mineral and organic matters, for this serum, were:

Amount of fixed organic matters	91.4
Amount of fixed mineral matters, ashes	8.6
	100.0

The most striking thing, above all, is not only that the serum of the blood of the horse contains less fixed organic and mineral materials, with a rotatory power very nearly equal to that of the blood of the ox and much lower than that of sheep's blood, but also that it contains between seven and eight times less mineral matters than either of the other two.

The serum of the blood of the horse therefore differs prodigiously from the two serums to which I had compared it; this amply explains at once the softness of the albuminoid atmosphere of the granulations and the rapid deposit of the globules.

Experiment thus verifies the fact that the molecular granulations occupy in the blood not only all the space occupied by the globules, but also all the space left free by their precipitation.

According to Charles Robin[3], agreeing with many authors, the blood of man, and that of the dog and the ox, behave like that of the horse when they are cooled down to a little below freezing; they remain liquid a sufficiently long time to enable the globules to be deposited, the leukocytes, according to Donne, forming a grayish layer upon the top of the blood globules. The supernatant liquid then forms the clot, losing its transparency when the temperature reaches 12° to 14°C (53° to 57.2°F) and it should be added that Robin, having observed the transparency of the supernatant layer separate from the globules under these circumstances, and which he called plasma, stated that it could not be filtered.

I regret that I had not had the time to verify these experiments, but the facts may well be regarded as true, being certified by Robin. They support the theory which I propose; the lowering of the temperature below zero (32°F) having the effect of singularly retarding the functions of nutrition of the anatomical elements (as it retarded those of beer yeast) ought to retard the coagulation of the albuminoid atmosphere of the microzymian molecular granulations.

The formation of the classical fibrin by whipping remains to be explained, and this is now very easy. It is the result of a simultaneous mechanical and chemical action. By the mechanical action, the layer of intercellular liquor which separates the molecular granulations is broken, while the granulations forcibly set free are united by their mucous albuminoid atmosphere. At the same time, the changes in the conditions of existence determine the allotropic transformation of the albuminoid substance which, coagulated as we have seen, contracts, still enveloping the microzymas.

Consequently, that which was diffused throughout all the volume of the blood is reduced to a relatively small volume occupied by the classical fibrin. And by the small volume of the fibrin obtained by whipping, a judgment may be made of the enourmous volume formed by the albuminoid atmosphere which surrounded the granulations of the flowing tissue at the moment of the venesection, as we have observed also in relation to the molecular granulations when separated from the alcoholized blood.

In the separation of fibrin by whipping, the globules remain entire, and I have explained that if the weight of the fibrin thus produced is less than that of fibrin obtained by washing the contracted clot, it is because to the free molecular granulations of the blood are added those obtained by the destruction of the globules with their envelopes.

Such were the facts upon which rested the experimental physiological theory of the spontaneous coagulation of the blood, when, in 1895, I communicated them to the Bordeaux Congress of the French Association for the Advancement of Science.

If there remained any doubt as to the value of the theory concerning the spontaneous coagulation of the blood as a flowing tissue, here is what must remove them.

The following experiments are the fruit of these considerations:

If the formation of the clot is actually the result of the spontaneous soldering of the mucous albuminoid atmosphere of the microzymian molecular granulations, and if in the presence of alcohol, diluted to a suitable degree, these atmospheres are condensed, the molecular granlations remaining independent of one another, what would happen if instead of alcohol the blood should be received into water?

The following is the experimental answer.

Coagulation of blood diffused in water.

At the moment of venesection, the blood is received into increasing volumes of distilled water up to half of its volume. The clot is formed in all cases. With small quantities, the globules do not appear to be altered, and the first serum has its ordinary appearance, but in proportion with the quantity of water being increased there arrives a time when the serum becomes colored.

Encouraged by these trials, one day in November, in Paris, the blood of the general venesection of a Russian sheep was received, as to one part into two volumes of alcohol at 36°, carbolated with two drops per hundred cc, and as to the other part into two volumes of distilled water carbolized in the same proportion.

The alcoholized mixture yielded the ordinary deposit of microzymian molecular granulations with the properties with which we are now familiar, and naturally there was no trace of a clot.

The aqueous mixture furnished something very different from a precipitate. Like the alcoholic mixture, the aqueous mixture had been made at 9 o'clock in the morning. Naturally, like the other, it was of a deep red color, since under such conditions all the haemoglobin of the blood globules had become dissolved. But at 3 p.m. the aqueous mixture was coagulated, the clot occupying the entire volume of the mixture, a volume three times greater than the blood clot without the addition of water. A little deep red liquid was already expelled; the next day the blood clot was not more contracted.

The experiment when repeated with ox blood gave the same results; the blood clot was formed in the entire mass, floating in a little deep red liquid.

The trembling and nearly transparent clots of the two experiments were placed to drain on a moist cloth of close texture. The cloths were soon covered with a mucous substance which a prolonged washing, with water slightly carbolized (1 drop per 100cc), then with alcohol at 25%, and again with water, did not completely decolor. At the end there remained a red false membrane which could be removed in a single piece from the damp cloth; such was the appearance and state of the fibrin of the clot formed under these conditions. Its quantity is perceptibly that which is isolated by the washing of the ordinary clot. In this condition this fibrin is not dissolved immediately in hydrochloric acid at 2/1,000, but like ordinary fibrin is a function of time and of temperature, and without first taking on the condition of a jelly like the fibrin from whipping.

If, then, the albuminoid atmosphere of the hematic microzymian molecular granulations is condensed under the influence of alcohol of proper strength, the same atmosphere is much more distended, three times more, in water, and the blood still coagulates in all its mass, the globules being destroyed.

The conclusion which results from this new series of experiments is that the physiological theory of the spontaneous coagulation of the blood, founded on the existence of a third anatomical element, the microzymian molecular granulations of the flowing tissue, sufficiently explains the facts.

It is time then to erase from the language of science the words *plasma, plasmine, fibrinogen* and *fibrinoplastic*, with which it has been encumbered.[4] There must also be erased from the explanation of the phenomenon the pretended influence of the globules, of calcareous or other salts, catalytic actions of contact, etc, not to speak of various occult influences.

An exact knowledge of the anatomy of the blood and of the conditions of existence of the anatomical elements will suffice.[5] But it is also necessary to understand in another way the meaning of what is called the coagulation of the blood. In truth, *the blood does not coagulate.*

The experiment proves it; it is the substance of the atmosphere of the third anatomical elements of the flowing tissue which, undergoing the allotropic change of coagulation, gives to the aggregate of the phenomenon the appearance of a total coagulation; but, as we have seen, it is only an illusion.

The supposed spontaneous coagulation of the blood is only the end of the first phase of the spontaneous alteration of the flowing tissue, just as the cadaveric rigidity marks the first phase of the spontaneous alteration of the muscular tissue.

But what is it that changes in a tissue? And what is the second phase of the spontaneous alteration of the blood and at what moment does it begin?

Second phase of the spontaneous alteration of the blood.
The first phase begins by the chemical alteration of coagulation of the albuminoid atmosphere of the microzymian molecular granulations, from which results the formation of the clot, the retraction of the latter and the expulsion of the lemon-colored serum. The globules have nothing to do with this phenomenon, as is incontestably proven by the experiment with the blood of the horse.

The second phase begins at the moment when the serum becomes colored with red, which shows that the change of the blood globules has commenced, their haemoglobin, more or less changed, being diffused in the serum.

The following experiment of Pasteur has shown what becomes of the globules in this change. The experiment was made in 1863,[6] five years after my verification of the hypothesis of germs in the air, when he had given up belief in the spontaneous generation of ferments, with the object of demonstrating that in the absence of germs the blood would not putrify because nothing living would appear in it. To understand this it should be remembered that Pasteur was a protoplasmist, seeing in an organism only proximate principles, admitting in it nothing figured, autonomically living, comparable to the figured ferments.

I take the recital of the experiment from a book of Pasteur's, published long after the microzymas had been discovered and the microzymian theory of the organization completed.

He commences as follows:

> " Let us examine into the interior of living beings, in good health, such or such of the materials which may be found there and examine them in the state in which life has formed them, in contact with pure air." [7]

In fact, with the assistance of Barnard, he poured the blood of a dog directly into a vessel, the air in which had been calcined. The receiver, sealed with the blow-pipe, contained thus one of the materials to be examined, obtained from the interior of the animal, and thus protected from the germs of the air. I now quote textually as follows what Pasteur thought he observed:

1) The blood does not putrify, even at the highest temperature of the atmosphere; its odor remains that of fresh blood, or takes on that of lye.

2) After an exposure of the flasks to 25°—30°C. (77°—86°F), during several weeks, nothing can be observed but an absorption of 2% to 3% of oxygen, which is replaced by a perceptibly equal volume of carbonic acid.

3) Under the circumstances in which the blood of the dog, exposed to the contact of pure air, does not putrify at all, blood crystals are formed with remarkable readiness.

4) In the first days of its being placed in the oven, slowly, at the ordinary temperature, the serum became colored a deep brown.

5) In proportion as this effect is produced, the blood globules disappear, and the serum and the clot become filled with crystals of a brown or red color. After some weeks not a single blood globule remains either in the serum or in the clot. After a longer interval

the whole of the fibrin may become collected in a single hyaline mass.[8]

Such was the experiment from which Pasteur concluded that, if protected from germs of the air, blood did not putrify at all; that is to say, was not altered by the action of any figured ferment, which in his opinion only the germs of the air could produce. I have elswhere shown that, assuming the technique of the experiment to have been accurately carried out, the observations made thereon were incomplete, and the interpretation of the results vicious in the extreme. I will revert to this hereafter; for the present I will only show that the facts of the experiment corroborate my own.

It is evident that, taken altogether, the experiment of Pasteur has confirmed that which the microzymian theory never fails to prove, namely, that every tissue and every humor when withdrawn from a living healthy animal, and absolutely protected from the germs of the air, necessarily alters and, consequently, alters spontaneously.

It demonstrates further that there are two distinct phases in the changes of the blood.

It is true that Pasteur did not stop an instant to consider the phenomenon of coagulation, but he observed that the serum, at first lemon-colored, became by degrees red, then deep brown, without insisting on the mechanism of the contraction of the clot and the expulsion and tardy coloration of the serum which marks the commencement of the second phase and which follows so exactly the consumption of oxygen which the blood contained that the author himself testified to the absorption of a small quantity of the oxygen in his flasks with a corresponding production of carbonic acid. During the second phase, in which the haemoglobin alters more and more, blood crystals are formed, at last the globules are destroyed and disappear while the fibrin which imprisoned them in the network formed by it contracts more and more.

This picture shows clearly that during the two phases the alteration is at once chemical and anatomical, ending in the destruction and total disappearance of the globules.

But to what cause did Pasteur ascribe such prodigious effects? In 1863 he also experimented on muscle flesh, imitating my method of investigation, replacing creosote with alcohol. He wrapped up a voluminous mass of flesh in linen, soaked in alcohol, and left it to itself.

"There will not be any putrefaction", he said, "not in the interior, because the germs of the vibrios are absent, nor externally because the vapors of alcohol hinder the development of the germs on the surface." Nevertheless, the author certified that the meat "became gamey in a pronounced manner." And why did it become gamey? Simply, he said,

> " because it is impossible at ordinary temperature to withdraw the interior of this flesh from the reaction of solids and liquids upon one another ... There will necessarily always be what are called actions of contact, which develop in the flesh small quantities of new substances, which add to the savor of the meat their own savor." [9]

Then at ordinary temperatures the same thing should happen with the blood as with the muscle tissues; there will be actions of contact, reactions of the solids upon the liquids. That is why the blood changes without putrifying as flesh becomes gamey without putrefaction.

There is some excuse for Pasteur and the academies with him to have been satisfied with these explanations, seeing that protoplasmism had been accepted as dogma. It was faith in this doctrine which caused the globule to be regarded as an organite, and the mass of the blood or the flesh to be regarded as a collection of proximate principles, and also, which is more serious, prevented Pasteur from seeing the microzymas among the results of his experiment, or, if he saw them, caused him to neglect them, as he had neglected the microzymas and even the vibrios in the gamey meat.

However it may be, it was by invoking actions of contact, and the reactions of solids and liquids, that Pasteur persuaded himself that the alterations of the blood and those of the flesh were not phenomena of putrefaction; that is to say, of fermentation. And this manner of regarding things on the part of the celebrated savant prevailed so generally that I was obliged to make an explanation of

CHAPTER SIX

the subject in reply to an assertion made by Balard in 1874. Servel, Estor's assistant, had presented to the Academy[10] a work verifying the fact that with absolute protection from atmospheric germs, the most diverse tissues could produce bacteria even in their interior, and cited other verifications made in Germany.

Now, Balard, who presented the work, took the occasion to say that blood is preserved without putrid fermentation and without bacteria in the experiment of Pasteur.[11] I replied to Balard, saying that the blood is one of these substances, and the egg is another, wherof the microzymas undergo with most difficulty the vibronian evolution.[12]

In spite of this the experiment of Pasteur was given as a preemptory demonstration that the interior medium contains nothing figured which could become bacterium by evolution in the substance of a tissue or of a humor; and it was asserted, in agreement with him, that the experiment proved at the same time that the bodies of animals were closed to germs from without.

Nevertheless, in a discussion at the Academy of Medicine, where, once more, I defended the microzymian theory, Pasteur took part in the discussion, maintaining his former conclusions, continuing even to deny the existence of the microzymas. It was then that I urged against him his own experiment upon the blood which is a demonstration against his own system. I said:

> "Do you affirm that the blood in which crystals are formed and the globules disappear is not altered? The globules of this blood are always destroyed and disappear; what then has destroyed them? Even the haemoglobin is transformed into crystals and we find in the liquid a swarm of microzymas ... these microzymas which you have neither seen nor noticed."

No longer invoking actions of contact, Pasteur said:

> "But these transformations are made under the influence of the oxygen of the air."

With regard to the presence of the microzymas, he admitted it in leaving it to be believed that I had stated that they became bacteria in his experiment.[13]

The presence then of the microzymas being acknowledged, and

the observation of the results of the experiment being completed, it mattered little after that that Pasteur continued to treat them as "creatures of the imagination," and that he explained the phenomena as the result of some influence of oxygen.

I had reason to hope that this avowal would open the eyes of my academical opponents and that they would acknowledge that they had been deceived. But nothing of the kind.

The acknowledgment exists nevertheless, and it only remains to prove that they are really the agents of the changes of the second phase of the phenomenon and the destruction of the globules.

In the first chapter, we showed that fibrin left to itself, covered by carbolized water, open even to contact with the air, is transformed into soluble products without the appearance of bacteria, leaving a residue of new microzymian molecular granulations without phenomena of fetid putrefaction; and we have also seen that the microzymas of these granulations were the ferments of the transformations.

On the other hand we have also seen that in fecula starch, the same fibrin liquefies this starch and makes it ferment, while its microzymas become bacteria; we have there two examples, in one of which the microzymas are active without evolution, in the other they undergo bacterian evolution.

The following experiment demonstrates that oxygen has no influence in the phenomenon of the destruction of the globules in defibrinated blood.

About 300cc of the blood of the ox, having added to it 50cc of a saturated aqueous solution of phenol, were immediately defibrinated and the blood carefully separated from the fibrin submitted to a current of carbonic acid, for the purpose of expelling the oxygen. The flask was closed and left to itself at the temperature of the month of June in Montpellier during one month, and afterwards in the oven at 30° to 33°C (86° to 91.4°F).

This blood did not undergo fetid putrefaction. The globules, very slightly altered in shape, remained whole during the first ten or twelve days. It was only on about the 15th of June that there appeared a great quantity of very fine molecular granulations, of

CHAPTER SIX

which only some rare examples had before appeared, without any trace of vibrios or of bacteria. The globules resisted for a long time further, and ended by disappearing. Here was an alteration without the presence of oxygen, wherof the microzymas could not be other than those of the globules.

The fibrin and the globules of defibrinated blood can then be destroyed by their own microzymas alone, without fetid putrefaction and without bacteria.

If in the experiment with the blood of the ox, defibrinated or not, no blood crystals were formed, it was because the haemoglobin of this blood is one of those which either do not yield them at all or do so with difficulty.[14]

Assuredly, if in spite of the changes certified to by himself, Pasteur came to the conclusion that muscle, flesh and blood were not liable to become putrid, it must have been because he firmly believed that ferments had for their only source the germs of the air, and that the protoplasmic system of organization was founded on rigorous observation.

And I venture to say that he knew he was in error and that, later, it was with design that he disputed the microzymian theory, being unwilling to confess that he had observed badly and had taken the wrong road.

Professor Joseph Béchamp reviewed the experiment on flesh as I had done for that of the blood. He repeated the experiment of Pasteur without using alcohol as an antiseptic, and in the centre of the piece of meat, there where Pasteur said that the germs of vibrios were absent, he found the microzymas in evolution and vibrios or bacteria. At the same time he found the tissue disorganizing.[15]

When I had led Pasteur to acknowledge the presence of the microzymas in the altered blood, I was anxious to make him confess the presence of bacteria in the interior of the gamey mass of flesh in his other experiments. But he refused, saying:

> "I do not know what you mean in speaking of one of my experiments on flesh."

The excessive role ascribed to the germs of the air by this savant and his pretended demonstration of the imputrescibility of organic matters in general when protected from the germs of the air have diverted science into a deplorable road.

He thus threw doubt upon a truth long since acquired; namely, that all natural organic matters, vegetable and animal, are liable to spontaneous change by a phenomenon of fermentation under the conditions specified by Macquer.

This truth must be re-established if we would seek to understand the real meaning of Pasteur's experiment upon the blood; to do this we must connect it with the introductory matter which precedes the first chapter of this work, which led up to the discovery of the real nature of fibrin, which was the point of departure for the discovery of the real nature of the blood.

I call to mind then that I proved how a solution of sugar or of any other proximate principle, or their mixtures, were changeable on contact with the air, owing to the ferments born of the germs of this air. Pasteur, who had preiously asserted the spontaneous generation of ferments, repeated my experiments and was convinced. Then he generalized and asserted that it would be the same in the case of urine and milk, just as sweetened yeast broth, which, when boiled, is not altered if creosoted or left to itself in calcined air.

Before the experiment on the blood or on the flesh, he had experimented on urine and on milk. As to fresh milk, he proposed, a priori, that it soured because of ferments born of the germs of the air, and that it was coagulated by the lactic acid which coagulated its casein.

But here we have boiled milk coagulating in calcined air without becoming sour, while vibrios appeared in it. He was surprised at this, but did not in any way seek to fathom the mystery, maintaining that in milk the germs of the air resist heating to 100°C and become vibrios, to which he ascribed the coagulation.[16]

I have narrated, in the introduction, how I applied the new method of research to milk and thence to various other tissues; similarly, I studied from the point of view of their chemical and

CHAPTER SIX

anatomical changes urine, birds' eggs, fruits which had become over-ripe, sprouted barley, frozen plants after a thaw and globules of beer yeast.

Now for the chemical and anatomical facts regarding milk, upon which I cannot insist too strongly.

The first phase of its alteration is the separation of the milk globules in the cream; this separation corresponds in an inverse sense to the separation of the blood globules in the cruor of the blood of the horse.

The second phase is the souring which precedes the formation of the clot. This souring corresponds to a fermentation which produces alcohol, acetic acid and lactic acid, the agents wherof are solely the microzymas proper to the milk, for at the moment when the clot is formed, whether the milk has been creosoted or not, the microscope discloses only these microzymas, which have become more readily visible. The vibrios or bacteria which then appear mark the anatomical phase of the phenomenon.

But for a complete understanding of the phenomenon of the alterations which occur in milk, it is necessary, as in the case of the blood, to recognize that after its issue from the gland its anatomical elements are no longer in their normal conditions of existence. Also, perfectly fresh milk does not contain lactic acid —contrary to the opinion of Berzelius—but it contains alcohol and acetic acid; it hence results that the production of lactic acid after the milking indicates a functional change in the microzymas of the milk, and this formation of lactic acid taking place without the disengagement of gas, especially of hydrogen, indicates in addition that the microzymas of the milk are different from those of the blood.

And since it relates to the phenomenon of coagulation and that the clot of milk has been compared to the clot of blood, it must also be recognized that the milk clot is not the coagulation of the casein by lactic acid.

The casein, in fact, is an insoluble albuminoid proximate principle, which exists in the milk in the state of a soluble caseinate. The acids, whether lactic or acetic acid, saturate the alkali, and the casein is precipitated. It results from this that that which is called coagulation of the casein in the milk, which is spontaneously altered, is the slow precipitation of the casein by the acids which render the milk sour.

With regard to the coagulation of boiled milk, where there is no souring; this is a phenomenon of another kind, in which the caseinates, the albuminates and the zymas of the milk modified by the heat take part. It is a zymastic action, which may be likened to coagulation by rennet, the zymas wherof has its origin in some functional modification by heat of the microzymas of the milk. And this functional modification of these microzymas is so certain that, if there be added to the milk quantities of creosote or carbolic acid sufficiently great to prevent the vibronian evolution of the microzymas, there will be no souring or coagulation of the milk. The albuminoid matters undergo other transformations and at last, if the action continues for a long time, at 30° to 35°C, the milk globules are destroyed, and the fatty bodies which they contain are set free.

The preceeding facts regard especially the milk of the cow and goat, which are casein milks.

The milks of the ass and of women do not contain casein, but they sour spontaneously without coagulating and yield no clot on the addition of rennet[17].

Normal human urine, creosoted, ferments without disengaging gas, producing alcohol, acetic acid and benzoic acid proceeding from hyppuric acid, while the epithelial cellules are destroyed and the microzymas evolute.[18]

The liver, plunged in carbolized water, produces, with the disengagement of carbonic acid, hydrogen and sulphuretted hydrogen, alcohol, acetic acid, and lactic acid, while its cellules are destroyed and its microzymas evolve and become bacteria.[19]

But the changes of eggs and of beer yeast are specially conclusive.

The egg of the bird is an organism whose function is to produce a bird. Donne, by vigorous jolting, destroyed this organism, mixing, in the shell, the yolk with the white and thus produced a kind of alteration which I studied.

The egg of an ostrich thus treated, at a temperature of 30°—35°C (86°—95°F), fermented and produced so much gas that the internal pressure became sufficiently strong to throw out a small part of the contents when a hole was made in the shell. The gases set free were carbonic acid, hydrogen and a trace of sulphuretted hydrogen.

When the gaseous disengagement ceased, there was no longer any sulphuretted hydrogen. All the vitellin globules had disappeared, and the microzymas were preserved with their form, without any trace of vibrios or other organized production. All the glucose of the egg had disappeared while the albuminoid matter had been preserved, the soluble being coaguable by heat. The products of fermentation were alcohol, acetic acid and butyric acid, showing that they had been produced from the lactate.

Here then was a fermentation, strictly defined, wherein the microzymas, like those of the blood, did not undergo vibronian evolution.[20] In order that the vitellin microzymas may evolve, other conditions are necessary.

The case of beer yeast is still more interesting, for it has to do with a living being reduced to a cellule, whose alteration and total destruction will throw a strong light upon those of the blood globule.

Consider an alcoholic fermentation of cane sugar for which a little more yeast has been employed than is needed for the complete fermentation of the sugar. The fermentation being accomplished, the yeast will be deposited in the fermented liquor and will be preserved there unaltered indefinitely, as if in lethargy, with its form and its properties.

This determined, let us take some fresh yeast, as it comes from the brewery, washed in distilled water to purify it from what it has brought away from the vat, and steep it in from three to four times its weight of creosoted distilled water to destroy the influence of germs of the air.

In this situation, so different from its normal condition of existence, at the temperature of about 30°C (86°F) and without any trace of air, it will for a long time disengage pure carbonic acid, producing at the same time a relatively great amount of alcohol, acetic acid and other products, preserving its form all the time. Evidently it has only been able to produce all these things at the expense of its own substance, of its contents, since its tegument at first remains whole. And if the process of alteration is allowed to continue, this tegument itself will disappear, and its microzymas will become free and vibrios appear.[21]

The following is the method by which the mechanism of the spontaneous destruction of the cellule of beer yeast can be most easily studied.

It is well known that yeast does not cause fecula to ferment. But what is not known is that it liquefies the starch of fecula and is completely destroyed in producing the liquefaction, leaving of its organism nothing but its microzymas, the soluble part of their content being left in the circumambient medium. The phenomenon lasts a greater or less time according to the quantity of creosote employed to destroy the influence of the germs of the air. If the quantity of ceosote is small, the microzymas undergo vibronian evolution, and if it is sufficiently large the microzymas do not evolve.[22]

But that is not all. Thus studied, the phenomenon of the spontaneous destruction of the cellules of beer yeast has enabled me to confirm the generality of the fact which I had long before observed in studying the microzymian origin of the vibrioniens.

While the globule of yeast is being destroyed and its microzymas set free and begin to undergo vibrionian evolution, several phases of this evolution are to be observed, which Estor and I have described from the commencement of our researches upon the liver, etc,[23] namely, at first the microzymas are scarcely altered in their size and form; then microzymas couple in the form of a figure 8, then chaplets of microzymas of from 3 to 10 and 20 grains, all of the same size; then vibrios properly so called; then bacteria often very large, motile or not; also the amylobacters of Trécul, either free or fastened end to end.

When the phenomenon is not checked by an addition of creosote or carbolic acid, all these productions may be seen at the same time in the field of the microscope. Now if without changing any of the conditions of the experiment the observation of it is continued, it will be seen that all the forms other than the single microzymas disappear successively; first the amylobacters disappear;

then new forms of smaller dimensions appear and disappear in turn, so that in the end there remain only swarms of motile forms scarcely differing from the original microzymas which had evolved.

Speaking then in the language of anatomy, we may say that the microzymas become vibrioniens by evolution; the vibrios, the bacteria, the vibrioniens in general, return to the microzymian form by an inverse phenomenon of evolution, the ultimate forms differing in little or nothing from the microzymas, the anatomical elements of the cellule.

It is thus directly demonstrated that a yeast globule, a cellule in general, in being destroyed sets its own microzymas free, and that these, if the necessary conditions are realized, become vibrioniens by evolution, which in the same apparent conditions, by an inverse phenomenon reproduce the microzymas.

So that as Estor and I have demonstrated in the development of the embryonic cellules of the fowl, and as I have demonstrated in the case of beer yeast and in the case of cellules which may develop in the mother of vinegar, the microzymas which are the commencement of all cellular and tissue organization are also their end, being, as we have seen, the end even of the bacteria.

That which is true of the microzymas of beer yeast is true also of the microzymas of all cellules, of all tissues, of both animals and plants. And this fact has been confirmed unwittingly even by those who deny the microzymas and who, to avoid naming them, have called them *punctiform ferments*; a microzyma or microzyma-producer at the beginning, microzyma at the end.

Such are the beginnings and the ends of a bacterium and of a cellule.

Thus all natural animal and vegetable matters, that is to say, organized as Bichat conceived them and defined their organization, the anatomical elements morphologically definite, are the only things living in them; yes, all these matters, from the highest in organization down to beer yeast, are spontaneously alterable from the moment that they are no longer in the situation of their natural conditions of existence, chemically and anatomically.

In insisting upon their chemical alterations, especially upon the production of alcohol, acetic acid, lactic acid and benzoic acid, with or without the disengagement of carbonic acid gas, etc, I wished to show that these alterations belong to the class of the best known fermentations, which assume a living figured ferment.

But even in the spontaneous alteration of beer yeast, alcohol and acetic acid are not the only products formed. I described others in 1864; on further studying these latter I have found succinic acid, a special gummy substance, a ternary, furnishing mucic acid, leucin and tyrosin, nitrogenous compounds whose formation bears witness that the albuminoids of the yeast contribute to the changes; later, others have been found equally nitrogenous. In extending these researches upon yeast to the spontaneous alterations of the flesh of the horse and fish, these researches have been verified by isolating similar or analogous products.

But since these spontaneous chemical alterations belong to the class of fermentations which presuppose the presence of a figured ferment, what is this ferment?

For if the beer yeast which causes sugar to ferment releases a part of itself which can be recovered from among the products of normal fermentation, it is not destroyed. It remains whole, its tegument preserving to its form, with its own anatomical element-microzymas.

On the other hand, when it produces alcohol spontaneously, without sugar, it alters, and is destroyed, as are destroyed the cellules and the organization of the blood, the flesh, or the liver. It is not then these cellules and these tissues which are the ferments of the spontaneous fermentations.

Pasteur sought in the altered blood the vibrio born of the germs of the air and, not finding it, concluded that there was neither fermentation nor even chemical alteration in it. There are, nevertheless, fermentations without vibrios and without cellules in which are produced alcohol and acetic acid.

In the first phase of the alteration of milk, for instance, and in that of eggs shaken within the shell, these ferments are precisely the microzymas, often the vibrioniens resulting from their evolution and microzymas which are the result of the destruction of the latter, for at a given moment, either at the commencement or at the end of the phenomenon, there is in the medium which is altering or of which the alteration is completed no production other than the original microzymas or the microzymas resulting from their destruction.

And this is not a gratuitous assertion, for I have experimentally proved that the microzymas of animal origin and those of the yeast are actually the figured ferments which produce, with sugar or fecula, alcohol, acetic acid, lactic acid, and by fermentation of the lactate of chalk, butyric acid. And the microzymas of the microzymian molecular granulations of the blood or those of the blood globules which belong to that class.

From all these experiments it results incontestably that the microzymas of living organisms in general, and those of the blood and the blood globules in particular, are anatomical elements and are themselves figured ferments; that is to say, that they are living and organized in the same manner as it is proposed that yeast is; as also are the vibrioniens which these microzymas may become by evolution, out of the same organized substance.

But the microzymas are living beings of an entirely special order, without analogy, on which I have insisted for a long time and again insist upon as crowning the demonstration that the blood is veritably a tissue.

And now what happens when this or any tissue alters?

First, it is no longer preserved in the state in which it exists and functions in the organism, in coordination with the functioning of all the organs and of their tissues.

It has then, as we have established for the albuminoid atmosphere of the hematic microzymian molecular granulations and for the coloring matter of the contents of the red globule, to undergo, owing to the change in the conditions of its existence, some chemical change in some of its parts. It is in short that its special anatomical elements change their form and their function to the extent of being destroyed and disappearing, leaving the microzymas as the only trace of their existence which, according to circumstances, do or do not undergo vibrionian evolution. And the anatomical change may be so rapid, as is well known to histologists, that one is obliged to take steps to preserve the integrity of the tissues. In fact, one or two minutes may suffice, after the blood has been shed, to render it impossible to demonstrate the third anatomical element.

We must conclude then that in all the experiments, including those of Pasteur, the chemical and anatomical alteration in the blood is the work solely of the microzymas, which, in certain conditions, do not become bacteria.

As to the question as to what order the chemical phenomenon belongs, it is now solved; since every chemical transformation of a proximate principle of organic matter, under the influence of a figured ferment, is called fermentation or putrefaction, it is evident that the spontaneous chemical alterations of the blood are the result of a fermentation or of a putrefaction without fetid products.

Assuredly, whenever the experiment on the blood shall be taken up upon a larger scale, even under the conditions of that of Pasteur, other products will be discovered besides those which I have pointed out, and among them I should not be surprised if alcohol should be found to be one of them.

The real meaning of the phenomenon called the spontaneous coagulation of the blood is now evident. It is the following:

The blood, being a tissue, is necessarily alterable itself, as are all tissues, and as are all natural organic matters, animal or vegetable. That which is called coagulation is only the first phase of its more complete change, which extends to disorganization and to the disappearance of its globules.

The phenomenon is solely the work of the microzymas, which, acting as ferments, effect the transformation of the proximate principles, and thereby the anotomical changes which end in the disorganization of the tissues and the cellules.

As I have said, the microzymas are living beings of a special order without analogy, upon which I have promised to insist afresh to give to this work undeniable certainty, and also to refute new errors which I have mentioned in a casual manner in a former work.[24] This will be the subject of the following chapter.

CHAPTER SIX

NOTES

1. (From p.134) It was, I think, in 1742, in his thesis entitled *Chylificationis historia*, that Bordeu maintained, among the original ideas which cause him to be regarded as one of the precursors of Bichat, that the blood is flowing flesh. In the seventeenth century, Amyot had already said that "the blood is engendered by the transmutation of some flesh which becomes a flowing liquid." (*Dict of Littre*) If an original sketch, later recognised as correct, is sufficient for the author to be historically regarded as the discoverer, assuredly Bordeu would deserve to be regarded as having discovered that the blood, like muscle flesh, is a tissue. But as observed by Babinet, "if the ancients have said everything, they have demonstrated nothing." Bichat also inserted the blood tissue among his twenty-one elementary tissues, next to his muscular tissues.

    But since Bichat, other savants have so done. In my time at Montpellier, the professor of Physiology Rouget taught that the blood, because of its globules, is a tissue; and I replied that, according to the ideas then accepted, the blood is no more a liquid or a tissue than was sweetened water holding in suspension globules of yeast.

    Today Ranvier also says that the blood is a tissue because it contains figured elements like the lymph. Doubtless the chief condition necessary for its being regarded as a tissue is for a product of an organism to contain some figured element, but that is not enough; according to the doctrine of Bichat it is also necessary to show that this element is living; and still that is insufficient; otherwise milk, saliva, and even urine and certain pathological serosities, spontaneously coaguable, would, like the lymph and blood, be tissues. I will consider this further in the last chapter.

2. The coagulation of the blood has been compared to the gelatinization of a solution of gelatine (Frey, *Traite d'hisologie et d'histochimie*, p.141), but a solution of pure gelatine heated in distilled water and sufficiently concentrated can be obtained absolutely limpid by careful filtration. On cooling, this solution forms a jelly, more or less consistent, perfectly limpid, not undergoing any other contraction than that produced by the lowering of the temperature. Nevertheless, in fact the gelatine has coagulated, for in the gelatinized solution it has become insoluble in cold water as it was before. But the comparison made by Dumas with the state of fecula in starch is more correct. In fact, in the transparent starch the fecula is not dissolved, it cannot be filtered. By cooling, after a long time, the starch undergoes a change in its appearance; it becomes more opaque and a retraction accompanied by expulsion of liquid can be observed. This happens because the fecula was not dissolved but simply enormously distended.

3. Robin, *Lecons sur les Humeurs*, p.59 (1871).

4. [The term *microbe* must also be thus erased; unless it be desired to retain it to denote mankind and other short-lived animals! Trans.]

5. It is because the conditions of existence of the anatomical elements are longer realized when the blood is preserved between two ligatures in the vessel which contains it that the coagulation of the flowing tissue is often long deferred. This explains the success of certain experiments of authors and, the most recent, those of Glenard (1875).

6. Pasteur, *Reserches sur la putrefaction.* C.R. Vol LVI (1863).
7. Pasteur, *Etudes sur la beere,* p.46 (1876).
8. *ibid,* p.49.
9. Pasteur, *Recherches sur la putrefaction,* C.R. Vol. LVI p.118-9 (1894).
10. C.R. Vol. LXXIX p.1270.
11. *ibid,* p.1272.
12. (from p.149) *ibid,* Vol. LXXX p.494. The note of Servel and my reply to Balard should be read with attention to have a clear idea of the state of the question in 1875.
13. *Bulletin de l'Academic de medecine,* 2d series, Vol XV p.679.
14. As to the assertion of Pasteur relative to the influence of the oxygen of the air; he knew that a long time before I had refuted this in advance (see *Les Microzymas* p.253, and following (1883), Chamalet, 60, *Passage de Choiseul*); it is there shown that the blood taken from the crual artery of a dog with the addition of a little of the saturated solution of creosote, submitted to a continuous current of common air is preserved arterialized, and the globules remain perfect a long time; these last, however, end by being destroyed so that the microzymas become free, often without the appearance of bacteria, and always without fetid putrefaction. When the current of air is replaced by pure oxygen the same thing happens and the crystals of blood are formed at between 24° to 26°C of temperature. It is on the contrary, in carbonic acid, that the blood globules of the dog are destroyed most quickly and the crystals are formed most readily between 33° to 40°C, always without fetid putrefaction.

    Later I showed that under the same conditions as that of the blood of the dog, the bloods of the ox, pig, fowl, and duck give neither crystals nor yet the soluble haemoglobin of the ox. *Bulletin of the Academy of Medicine,* 2d series, Vol XVII, p.225 (1887).

    I will add that under the prolonged action of the current of air on the blood of the dog I have found that the quantity of normal urea was increased. This statement should be verified.
15. C.R. Vol. LXXXIX p.573.
16. His memoir (*Ann. de Chimie et de Physique,* Vol LXIV pp.58-63) shows the efforts made by Pasteur to convince himself that the germs of the air are the sole origin of the vibrios.
17. *On the histological constitution and the comparative chemical compsition of the milks of the cow, of the goat, of the ass and of woman* and *On the spontaneous alterations of milk and on the changes which heating produces in it.* (Chamalet, Passage de Choisel, 60 Paris.)
18. C.R. Vol LXI p.374 and *les Microzymas,* etc., p.713.
19. C.R. Vol LXXV p.1830.
20. C.R. Vol LXVII p.523
21. For details and developments see C.R. Vol. LVIII p.601; *Sur les fermentations par les ferments organises* (1864).

CHAPTER SIX

22. *Ann. de chimie et de physique*, 4th series Vol. XXIII p.443 and *Sur la nature et l'origine des ferments* (1871).
23. *C.R.* Vol. LXVI p.421 and p.859 (1868).
24. *Microzymas et Microbes* etc. Chamalet pub, Paris 60 Passage de Choiseurl 60.

CHAPTER SEVEN

Proof that the blood is a flowing tissue and therefore spontaneously alterable.

Pasteur and the germs of the air.

Robin and the alteration of the blood.

Microzymas and spores of schizomycetes.

Microzymas and micrococcus.

The microzymas and the circulatory system.

Comparison of the microzymas of the blood, the circulatory system, and other tissues.

Autonomy of the microzymas.

The demonstration that the blood is a flowing tissue and therefore spontaneously alterable rests entirely upon the discovery of the microzymas, individual living organisms which had been previously unsuspected, but were found to be figured elements in all the parts of every living organism, in every cellule of that organism, *ab ovo et semine*, during the entire duration of its development and its existence in a condition of perfect health. This discovery has furnished the demonstration that all the tissues and humors are spontaneously alterable, because they contain, inherent in themselves, the agents of their alterability, the microzymas, which by evolution may become vibrios or, in certain fixed conditions, bacteria. All of this has been disputed and even denied.

It is this principle, so comparable to the conception of Bichat regarding the existence of autonomically living anotamical elements, and so entirely opposed to the doctrine of a living matter without living figured elements, called protoplasm or blastema, which I have

CHAPTER SEVEN

had to oppose and still oppose, in order to confirm the fact that the blood is really a tissue and as such is spontaneously alterable.

The following is a summary of the starting point of both the dispute and the denial.

The fact of the spontaneous alterability of organic matters under the conditions which Macquer specified was held in science to be an incontestable truth. I have described, in the first chapter, how this belief had been so generalized that the spontaneous alteration of all proximate principles (even that of cane sugar) was accepted. But, as I have demonstrated, it is only through the action of germs of the air, whose existence (even notwithstanding the hypothesis of Spallanzani) was denied, that this alteration which appeared to be spontaneous occured.

At the same time that I demonstrated that Macquer was right as regards plant and animal tissues and humors, I showed that of the three conditions specified by Macquer (suitable humidity, a certain temperature and momentary contact with the air), only the first two were essential. The air and its germs could be entirely suppressed.

Pasteur proposed the spontaneous alterability of organic matters in general, and explicitly asserted that ferments, beer yeast, lactic yeast and vibrios were spontaneously born of the albuminoid matter of the broth of sweetened yeast; Pasteur, having repeated my experiments, was so convinced that germs do really exist in the air, and that he had been mistaken, that he declared that the sole origin of the ferments, vibrios included, was these germs which he had previously disregarded, and that consequently these germs were without exception the first cause of the spontaneous alteration of all organic matters.

He experimented for the purpose of proving that without the germs of the air, unputrefying corpses would accumulate upon the earth, and he even calculated the consequences of such an accumulation. To subsequently deny the microzymas and contest the consequences of their discovery was a step which, later, he did not fail to make. In fact his own experiments on milk, boiled urine, blood and raw meat were made in the ardor of this new conviction, for the purpose of combatting the doctrine of spontaneous generation by the same weapons I had employed, and not against the microzymas which I had not yet named. It was only in 1876 that Pasteur began to deny and dispute the facts of

THE BLOOD AND ITS THIRD ELEMENT

the microzymian theory which had been nearly all published in 1871, and since 1874 had been verified and confirmed both in France and abroad. But although verified and confirmed, the theory was also subject to various interpretations; for the sake of history and comparison, it is desirable that these interpretations should be known.

Charles Robin was the first to speak of microzymas as being things which may become bacteria by evolution, but the way in which he understood the existence of the microzymas in the animal body needs to be mentioned.

He accepted without difficulty the two meanings attributed by Pasteur to his experiment upon the blood: firstly, that the blood does not change of itself, and secondly, that the bodies of animals are closed to germs from without, and consequently, that within the body there is nothing which could become bacteria.

Robin even asserted that Pasteur had proved positively and beyond question that the human body is absolutely closed to penetration by bacteria. Nevertheless, the observations of Davaine, Rayer, Coze and Feltz, etc, had demonstrated that in certain diseases bacteria appeared in the blood.

Unwilling to admit that the microzymas existed in the blood as anotomical elements, they said that either bacteria are the results of a spontaneous generation of the microzyma, passing into the state of bacteria, or that the microzymas reach the blood by penetration in the same manner as granules of dust, etc.[1] This alternative exposed the perplexity of this position.

In fact, Robin was such a protoplasmist, after a fashion; an anotomical element such as the microzyma, passing, as he said, into the state of a bacterium, among the ordinary anotomical elements, which he knew so well, disarranged all his ideas. But with the loyalty of an impartial man of science, he did not hesitate to class the microzymas in the same category as the bacteria; thus in an article, he asked: how do the microzymas come into the living organism? It was doubtless the perplexity of which I have spoken above which caused him to compare the microzymas to the micrococcus of the botanist, Hallier or to identify them with the *bacterium punctum* of Ehrenberg.[2]

Two years later an honest savant of Switzerland stated as follows:

CHAPTER SEVEN

" It is within my knowledge that it was A. Béchamp who first regarded certain molecular granulations, which he named microzymas, as being organized ferments, and formulated the three following propositions, based upon researches which he had pursued jointly with Estor:

1. In all the animal cellules which have been examined, there exist normal granulations, analogous to those named microzymas by Béchamp.

2. In the physiological condition, the microzymas preserve the apparent form of a sphere.

3. Outside the body, without the intervention of any foreign germ, the microzymas lose their normal form. They begin by becoming associated in chaplets, of which a separate genus has been established under the name of *torula*; next they become lengthened, so as to resemble bacteria isolated or associated ... "

and he added:

" It is evident that the subsequent researches of Billroth and Tiegel are in their results only the confirmation of these three propositions."

Then, experimenting on the pancreas of ruminants and freshly killed dogs, he declared that there were always found the same molecular granulations, exhibiting brownian movement, which became vibroniens by evolution. These molecular granulations, he exclaimed,

"... are evidently the microzymas of Béchamp, the coccos of Billroth and, without hesitation, the *monas crepusculum* of Ehrenberg."[3]

Later, Nencki, in collaboration with Giacosa, confirmed our observations generally, working upon the same tissues as we had done, but was unwilling to class the microzymas as anatomical elements to the extent that when the bacteria and vibrios were no longer thought to be animals, they were regarded as plants, under the name *Schizomycetes*. He came at last to hold that the microzymas are the spores of these infusorial plants.

Thus the facts were verified and confirmed in every sense; they exist in all the parts, down to the cellules of every living organism, *ab ovo et semine* of figured ferments and are capable of becoming bacteria; but instead of regarding them in such situation as autochthones (aborigines) they were regarded as being either the fruit of spontaneous generation according to one of the suppositions of Robin, or as foreigners under the names of *bacterium punctum, monas crepusculum, coccus, micrococcus*, pointed microbe, and finally of spores of *schizomyctes*.

Nevertheless, if the microzymas are not what I contend they are, autonomous anotomical elements, the alternative stated by Robin remains; spontaneous generation or penetration! But then what becomes of the dogma of closure, and that of non-putrefiability? These will be abjured rather than admit the microzymas to be essential anotomical elements! In fact, Cornil declared as follows, in 1886:

> " Pasteur has abundantly demonstrated that our tissues and interior media, like the blood, contain no microorganisms, no more than the urine, except such as have been introduced from without, and the experiments of our illustrious colleague have been confirmed in all countries."[4]

Then Cornil, continuing to deny the facts I had advanced, but admitting the views of those who believed in parasitic microzymas, exclaimed:

> "Messrs. Nencki and Giacosa regard the word microzymas as the synonym for *micrococcus*; if this synonymity be admitted, if the microzyma is merely a genus of the Schizomycetes, the word microzyma ought to disappear and the whole doctrine of Bechamp will vanish."[5]

But after Pasteur's admission of the prescence of microorganisms in the altered blood of his experiments, it was more than ever necessary to get rid of the annoying word microzyma; therefore Cornil went to Germany to call Nencki to the rescue. He replied (according to Cornil):

> " The microzymas of Béchamp are in my opinion either the cicrococcus or spores of bacteria and you are right in saying that for me the microzymas of Béchamp are spores of Schizomycetes."[6]

This reply of Nencki was communicated by Cornil to the Academy of Medicine.

CHAPTER SEVEN

If Cornil was satisfied, he was satisfied with very little, since his correspondent could not go back on his interpretation of ten years before. In fact, the matter in question was not one of synonymy and interpretation, but of a principle disputed and of facts denied by himself, following Pasteur. This principle and these facts—did Nencki deny them? That is the question. The principle disputed is the following, just as I had enunciated in a letter to Dumas in 1865:

> "Chalk and milk contain living beings already developed, a fact which observed directly is also proved by this fact: that creosote employed in a non-coagulating dose does not prevent the milk from clotting later; nor the chalk from transforming without outside help, sugar and fecula, into alcohol, acetic acid, lactic acid and butyric acid."[7]

The following year (1866) I gave the name of microzymas to the living beings already developed in the chalk and milk, so as to mark the fact that they were figured ferments. It will be seen that this bringing together the chalk and the milk was intentional on my part.

It was the principle, derived from experiment, that creosote which hinders the proximate principles from altering on contact with a limited quantity of air does not prevent natural organic matters from being altered in fermenting, which was disputed; and it was the presence of the microzymas, agents of these spontaneous fermentations, and their capacity to become bacteria by evolution, which was denied. But Nencki admitted both the principle and the facts; he had even avowed that Billroth and Tiegel had only confirmed the facts.

After that, it is of little importance that they have said in turn that the microzymas are the *bacterium punctum*, the *monas creusculum*, or spores of bacteria called *schizomyctes* after having been regarded as animalcules. I remark only that these various appellations prove merely that they do not know what to believe; but we shall see at the end of this chapter that the name microzymas has been well chosen, and that they are what they have been said to be—anatomical elements and living beings of a category not before suspected, and without analogy.

Meanwhile, the principle of the demonstration that the blood is a tissue whose change by fermentation, outside of the vessels, is spontaneous, as is that of every other tissue outside of the body, is certain, both by the acknowledgment of Pasteur and by the declaration of Nencki obtained by Cornil. But if the principle is recognized, can it be asserted that the fact that the blood is a tissue has not been sufficiently proved? It is necessary to insist further.

I have already remarked that it is not enough that figured elements exist in a humor to entitle the humor to be regarded as a tissue. In the order of the ideas of Bichat concerning elementary tissues, it is necessary to prove that these figured elements (i.e. having a certain form), regarded as anatomical elements, are not really living; this is what I began doing; but even this is not enough. It must further be shown that, as in tissues generally, these elements, almost in contact, are separated and yet connected among themselves by an intercellular substance in such a manner that the smallest mass of the complex tissue contains them.

If the blood were a homogenous liquid holding the microzymas in a state of isolation from the fibrin, that is to say, naked, in suspension with the globules, they would be separated and deposited despite the movement of the blood, because they are of greater density that it, in the same manner that rivers charged with argilacious mud deposit it despite the motion of the water. But the blood does not contain the microzymas in a naked state, but rather surrounded by an atmosphere of special albuminoid matter; in short, the blood contains the microzymian molecular granulations, and the albuminoid atmosphere, which is mucous, hyaline and swollen, gives to these granulations a density differing very little from, and perhaps the same as, that of the intergranular and interglobular substance which connects them, in such a way that both the molecular granulations and the globules pervade the entire mass of the blood at the same time.

The structure of the haematic-microzymian molecular granulations is precisely that which is needed to make the blood, with its globules, a tissue. It is because of their mucous atmosphere, which swells enormously, that these innumerable microzymian molecular granulations occupy in the blood the entire space not occupied by the globules and the thin bed of the intergranular and interglobular liquid substance; and it is due to this special viscosity that the swollen mucous atmosphere of the microzymian molecular granulations, and also to the mechanical obstacle which these present, that the globules remain

CHAPTER SEVEN

uniformly disseminated and are not precipitated during coagulation outside of the vessels before the production of the clot.

As for the special case of the blood of the solipedes, it is due to the great difference between the density of the globules of their blood, and to some peculiarity of the mucous atmosphere of their microzymian granulations, connected with the lower density of the intergranular liquid.

The demonstration that the blood is a tissue, and a flowing tissue, follows from the relationship between its three anatomical elements and the intercellular liquid substance special to each species.

But the blood, as a tissue, belongs to a special anatomical system of organs of which is the content. If it is true that the various anatomical systems are differentiated by their microzymas as they are by their form and structure, must it not be the same with the circulatory system? In fact, that is the case.

The microzymas of the vascular system, container and content, are different from those of the other anatomical systems.

I have proved this proposition in the comparative study of the decomposition of oxygenated water by the microzymas of various animal tissues, and I then extended this study to that of the microzymas of various plant tissues.

The results will be found in the following tables and have been obtained as follows:

Into a graduated tube, over mercury, are introduced several cubic centimetres of non-acidulated oxygenated water of known standard. The tube is then reversed, and one c.c. of microzymas in cake, enveloped in silk paper, is introduced into from 3 to 5 vol. of water, oxygenated to 10 to 12 vol. of oxygen, and the rapidity and the volume of the oxygen set free over 24 hours are noted. As mercury by itself can set free oxygen from oxygenated water, a tube having the same volume of this water serves as a control.

A similar tube receives the dust of the laboratory introduced under the same conditions as the microzymas.

TABLE I

Microzymas and tissues obtained from the blood of the regions of the circulatory apparatus.

			oxygen set free (c.c.)
Microzymas of	fibrin (sheep or ox)		23
"	"	blood (not defibrinated)	25
"	"	defibrinated blood (i.e. of the globules)	20
"	"	the lungs of a sheep	27
"	"	the lungs of a dog (the lung was antracosed)	29
"	"	sheep's liver (the liver had been drained)	21
"	"	sheep's liver (with bacteria, the microzymas having partially evolved)	22
"	"	spleen of a dog	10
"	"	heart muscle of a dog	12
Chopped muscle, washed, of heart of dog			8
Microzymas mixed with bacteria, of human urine			14
Controls:	Tube of mercury only		0.5
	Tube containing dust from laboratory		0.5

CHAPTER SEVEN

TABLE 2
Microzymas & various tissues not belonging to the circulatory system.

		oxygen set free (cc)
Microzymas	mixed with bacteria of human saliva	2.0
"	of the gastric glands of a calf	0.4
"	gastric, of a dog obtained by means of a fistula, isolated from the gastric juice	6.0
"	Pancreatic, of a dog	3.0
"	Pancreatic, of an ox	3.0
"	Pancreatic, of brain of a dog	2.8
Pulp of a dog's brain		1.3
Crystallin of ox		0.3
Cornea of ox		2.0
Vitreous humor of ox		0.7
Ciliary processes of ox		2.5
Sheep's periosteum		0.8
Sheep's bone		0.8
Sheep's hoof		0.6
Costal cartilage of calf		0.6
Nails of man		0.4
Vitellin microzymas of fowl		3.0

TABLE 3

Microzymas and various tissues of plants

	oxygen set free (c.c.)
Microzymas of sweet almonds	24.0
Tissues of sweet almonds finely divided (cotyledon and embryo)	8.0
Microzymas of beer yeast (isolated by braying)	3.6
" " " " (another sample)	11.0
Beer yeast, quite fresh pure	22.0
Green leaf, brayed	3.8
Crushed (brayed) yellow petals of a lily	2.0
Red petals of a cruciferous plant (brayed)	1.0
Pollen of an iris (in 20 minutes)	14.0

A comparison of the results of these three tables is very instructive.

From a comparison of the first two, which relate to the tissues of animals, it is seen that the microzymas of the circulatory system, including those of the urine, are those which decompose oxygenated water with the greatest energy, setting free the most oxygen. Of these, it is the hematic microzymas and those of the lung and liver which are most active; and these are the organs which are most directly concerned in the circuit.

This I wish to make especially clear, to demonstrate that the circulatory system is differentiated from the other anatomical systems by a special property of its microzymas; a property so special that one might almost think that the other tissues owe their like power only to the hematic microzymas which they retain. But this cannot be, for the microzymas of the thoroughly drained liver are as active of those of the blood.

The results of the second table are still more significant, for one cannot suppose that any hematic microzymas can be present among the vitellin nor yet among those of the saliva and urine. And if there remained the least doubt, the result of the third table must remove it, through consideration of the action of the amygdalic microzymas and that of those of beer yeast, which further proves that differences of the same kind are presented by the microzymas of the different plant tissues.

The microzymas of the vascular system differ from the microzymas of the other anatomical systems with regard to their power of decomposing oxygenated water. This is also to be seen from the above mentioned observations of Thénard when correctly interpreted. These differences are seen to be still greater when we study comparatively the physiological functional aptitudes of the various anatomical systems in man and other animals.

For instance, while the pancreatic and gastric glands of the dog and of ruminants are endowed with like functional properties in digestion, this is not the case with the salivary and parotid glands of man and those of dogs or horses. The salivary and parotidian microzymas of man powerfully liquify and saccharify the starch of fecula; the same microzymas of the dog or horse liquify only slowly and do not at all saccharify the same starch. Thus the zymas secreted by the microzymas of the same gland in man and in other animals is essentially different. Morphologically identical, these microzymas are functionally different, and I am certain that the more these are studied the more reasons will be found for differentiating the microzymas of the microzymian molecular granulations of the blood of the various species of animals and those of their globules, as I have differentiated the haematic microzymian molecular granulations.

And the microzymian theory of the living organization explains why this should be; it is because the microzymas of each species are autonomous in it and are, *ab ovo*, what they should be in order that each species should propagate itself, develop itself, preserve itself, and after death, thanks to oxygen, that each individual should undergo that total destruction which reduces all substances except the microzymas to the mineral condition.

If they were not anatomically autonomous, why should they differ and be functionally various in species and in their anatomical systems? I have answered this question in the past, and have been met by only bald denials. It is now worthwhile to introduce new considerations to convince those whom the assertions of Cornil and Nencki might yet lead astray.

NOTES

1. Robin, *Lecons sur les Humeurs*, p.225, 1874.
2. *ibid loc cit*, p.250 (1874).
3. Dr M. Nencki, *Ueber die Zersetzung der gelatine* and *Des Eiweisses bei der Faeulniss mit Pankreas*, p.35 Berne Dalp'sche Buchhandlung (1876).
4. *Bulletin de l'Acad. de Med.* 2nd series, Vol XV p.259 (1886).
5. *ibid*
6. Cornil did not say spores, but genus of Schizomycetes, which though very different is erroneous none the less.
7. *Annales de Chimie et de Physique*, 4th series Vol. VI p.248.

CHAPTER EIGHT

The microzymas and bacteriology.

Ovular and vitellin microzymas.

Microzymas and molecular granulations.

Geological microzymas.

Biological characteristics of microzymas.

Microzymas and their perennity.

Microzymas and pathology.

Phagocytosis.

Microzymas and anthrax.

Microzymas and disease.

Microzymas and microbes.

Microzymas and the individual coefficient.

Microzymas, life and death.

Microzymas and health.

Microzymas, blood and protoplasm.

Conclusions.

To place beyond dispute the autonomy of the microzymas, it is first necessary to consider the facts and observations which prove that the existence of the microzymas as living beings has not been suspected by those naturalists who have studied the infusoria, nor yet by the anatomists who have studied the cellules and tissues.

Demonstration that the microzymas are living beings, belonging to a category of their own, and having no analogue.

Let us first get rid of the hypotheses that the microzymas are either the *bacterium termo*, the *monas crepusculum*, the *micrococcus*, or the spores of bacteria.

It is to be borne in mind that I gave the name of *microzyma* at first to the geological figured ferment of the chalk of Sens and other calcareous earth, and that I have discovered this ferment in other calcareous rocks, always of a spherical form, very brilliant, exhibiting brownian movement and smaller in size than all the vibrioniens described by authors.

Ehrenberg described (in the chalk) the remains of fossil microscopic organisms called *polythalamies* and *nautilites*, but makes no mention of either *monas crepusculum* or *bacterium punctum*. In fact, none of the microzymas can be confused with those described by Ehrenberg under those names. The microzymas are even smaller than the *bacterium termo*, the smallest of the known infusoria, the first term of the animal kingdom, according to Felix Dujardin.

Nevertheless, the microzymas had been seen in cloudy infusions of vegetable and animal matters, but they were thought to be "the active molecules of Robert Brown", i.e. they were thought to be molecules having staggered or scintillating movement without changing their place, in other words "brownian movement," and no further attention was paid to them.

In fact, the microzymas are neither the *bacterium punctum*, nor the *monas crepusculum*, nor even the *bacterium termo*, which is much smaller than they. It will be sufficient to establish this fact by referring to the description of these monas, given by Dujardin in his *Historie Naturelle des Zoophytes*, pp.215 and 279.

On the other hand, if these bacteria, these monads, these micrococci, belong to determined species, it is contrary to the evidence of natural history to regard them as capable of being transformed into other genera and species of vibrioniens, as we see the microzymas produce them by evolution.

The suggestion that the microzymas are the spores of schizomycetes is also untenable for the following reasons:

A spore is an egg, if according to the old view, the bacteria are

CHAPTER EIGHT

animals, and search has been made for the eggs of bacteria; or it is a grain, if according to the new creed the bacteria are vegetable. But egg or grain, a spore cannot multiply itself as the microzyma does, and cannot therfore be the same thing.

Take for example the microzymas of the ovule in the Graafian vesicle in the fowl, and the microzymas of the vitellus of the mature egg. In the ovule there are ovular microzymas, and in the vitellus, vitelline microzymas. At a given moment there are, say, a milligram of microzymas in the ovule, and there are two or three grams dried at 100°C (I have isolated and weighed them) in the vitellus.[1] They have then multiplied prodigiously during the development of the vitellus.[2]

So much, then, for the anatomical analysis of the egg of the fowl. Chemical analysis shows that the elementary composition of the ovular microzymas is not the same as that of the vitellin, the former, as will be seen, being less carbonized. Evidently, their composition changed in the process of multiplication.[3]

Chemical analysis has further demonstrated that the vitellin microzymas of several species of birds differ from those of the fowl in their composition and especially in the properties of their respective zymases.[4] This accords exactly with the microzymian theory, for it is evident that the microzymas are what they should be, so that the egg can produce the bird, its tissues, and all that pertains to its future being. It has been demonstrated that during the development of the being, parallel with the anatomical development by the multiplication of the microzymas, there is a functional development of these, so that in each anatomical system they become that which they successively are in the embryo, in the foetus, and the adult.

If the hypothesis that the microzymas are the spores of bacteria were true, it would be necessary that there should first have existed in the circumambient atmosphere as many species of these spores as there are species of animal and vegetable ovules; next it would be necessary for these spores to penetrate as far as, and into, the ovule, and they would there have to multiply so as to fill up the vitellus of the egg of the fowl.

I need go no further, for there are still otherwise enormous difficulties, when we take into consideration the microzymas of the developed being, which are so different from the embryonal and foetal microzymas! But it now lies with the opponents of the microzymian theory to demonstrate the existence of these spores and their penetration as far as, and into, the ovules, and their multiplication. We have thus discarded the hypothesis opposed to that of autonomy. It is also discarded by the following consideration.

Shortly before Pasteur's admission in 1886 of the presence of the microzymas in the altered blood of his experiment, he had, for the purpose of denying them, asserted that the microzymas were the molecular granulations "which we all know." This was to his confreres at the Academy.

Histologists and pathologists knew of microzymas and represented them by a stippling in their diagrams of tissues. But their name even betrayed their opinion that they were neither organized nor living; in effect, the qualification of 'molecular' was intended to indicate that it meant only small collections of some sort of matter; thus they were described as white, gray, minerals, fats, albuminoids, etc. They were even described as posessing the brownian movement; nevertheless, before the discovery of the microzymas, no one thought of connecting them either with the *bacterium punctum* or the *monas crepusculum*. They were connected with anatomical organisms as being the remains of tissues, or destroyed cellules, or as amorpyuous matter; no one dreamed of making them come from outside. No consideration of the anatomical molecular granulations had anything to do with the discovery of the microzymas, except, as I have shown above, purely chemical considerations.

No, the molecular granulations are not the microzymas. And from the time of our first note, Estor and I have stated that the microzymas exist only among the anatomical objects which in histology are called molecular granulations. We held the microzymas to be autonomous anatomical elements. A more careful anatomical analysis enabled me to demonstrate that there exist naked microzymas and microzymas in the condition which I have termed microzymian molecular granulations.

Thus is disproved another gratuitous and erroneous assertion!

I return now to the microzymas. I had described them from the beginning as being chemically and physiologically figured ferments, producers of zymas, which are called soluble ferments, and were placed in the same category as the figured ferments which are insoluble. Biologically, I distinguished them as being able to become vibrionien by evolution, a fact which we have seen to be verified in every sense. But in the experiments on spontaneous alterations, or fermentations, wherein microzymas become bacteria, we have seen that these were destroyed and that vibrioniens more and more minute appeared in their place, so that at last there remained only of these bacteria the forms nearest to the microzymas. In the same manner that by their destruction the cellules set their microzymas free, the bacteria in their complete destruction reproduce microzymas similar to those of the chalk, and we will now see how that is.

In the experiments on the spontaneous alterations of natural animal matters, the substances which in a chemical sense are termed organic and which result from transformations by fermentation under the influence of the microzymas, both before and after their vibrionian evolution, and with or without the setting free of gas, are never entirely destroyed. In other words, they are not reduced to a mineral condition, e.g. carbonic acid, water, nitrogen, etc. For such destruction, oxygen is necessary, under conditions which reproduce those realized in geological epochs.

When I had discovered the microzymas in the chalk and other calcareous rock, and became convinced that they were not dependent on atmospheric germs, I asked myself if they were not the living remains of organized beings which had disappeared in geologic times.[5] This hypothesis was verified in the following manner:

A kitten was killed and buried between two beds of pure carbonate of lime and left in a cylindrical glass vessel, covered with a small quantity of paper so that the air had free access to it, but dust was excluded.

It was left in this state for seven years, at the end of which every part of the body, except some fragments of bone, had disappeared. The carbonate of lime was perfectly white, so complete had been the destruction. Under the microscope, nothing was to be seen in the upper layers of the carbonate except microscopic crystals of aragonite of this carbonate; but in the beds adjacent to the place, and underneath, where the kitten had been, there were crowds of glittering motile microzymas, such as are to be seen in the chalk of Sens.

And with this kind of artificial calcareous rock, containing the microzymas of an animal of the present day, I was able to repeat the experiments on fermentation which I had made with the chalk of Sens and with other calcareous rocks, both lacustrine and marine.[6] Such was the first experimental verification of the hypothesis that the microzymas of the chalk and calcareous rocks are the organized remains, still living, of beings which lived in the geological ages of the earths to which those rocks belonged.

I have said that the microzymas of the artificial chalk were the microzymas of an animal of the present epoch, but this needs some modification in terms to be quite accurate.

There were the microzymas of the bacteria which the normal microzymas of the animal had first become by evolution. By fresh experiments I have learned that the microzymas of an entire body, or of the liver, the heart, the lungs or the kidneys, become bacteria in the first phases of the phenomenon; these then disappear, becoming again microzymas, while the rest of the matter already transformed is, under their influence, and with the access of air, reduced to the mineral state, i.e. carbonic acid, water, nitrogen, etc.[7] And I have demonstrated that whereas in the climate of Montpellier seven years were required to accomplish this, a much longer time would be needed in a colder climate, so that in a climate such as that of the Obi valley, centuries would be required.

It was then a legitimate conclusion that the microzymas of calcareous rocks, clays, and marls, in short, of all the rocks which contain them, are the organized and living remains of animals and plants of past geological epochs. These beings were histologically constituted, as are the beings of our epoch, so that when they died, their microzymas became bacteria by evolution, and the microzymas, geological ferments, of these rocks, are those of these bacteria destroyed in their turn and reduced to their microzymas.

It is not surprising then, that having long pursued the anticipated consequences of the hypothesis now verified, I have demonstrated the presence of the microzymas in the earths of Herault and Gard, in cultivated lands generally, in moor lands, alluvials, in water, in the dust of the streets, where they are to be found in crowds—often still in the condition of bacteria, proving that like those of the calcareous

CHAPTER EIGHT

rocks, they are energetic ferments. And already, prior to 1867, I had made known their role in the soil in agriculture.

These researches led to a result of very great importance; it was the demonstration that what were and still are called germs of the air are essentially nothing other than the microzymas of beings which have lived, but have disappeared or are being destroyed before our eyes. In fact, by precise experiments, I have proved that the microzymas of the air are ferments of the same order as those of the chalk and rocks, and also those of my experiments with artificial chalk; only, varying with the places, the circumambient air may, along with these microzymas, contain conides of lichens, spores of mushrooms, bacteria and everything else that the wind can disperse.[8]

There is then no panspermy such as that which Charles Bonnet had invented, nor that which Spallanzani and Pasteur (after me) had proposed. In short, there are no pre-existing germs. There exist in the surrounding air only the microzymas of former beings which have disappeared and are disappearing with the things which the wind scatters.

Let us reflect firstly that the species of microzymas are as numerous as the species of eggs, seeds, and spores of the various species of animals and plants, and secondly that there are, in each animal and vegetable organism, already developed or in process of development, microzymas as numerous as there are anatomical systems and organs, tissues and special cellules in these organisms. It is then easy for us to understand that the species of atmospheric microzymas are present in enormous numbers. One can also understand the very great number of changes which these microzymas may cause when they enter a fermentescible medium in which they can multiply, and either evolve in it or build in it cellules or a mould.

If then, as I have demonstrated experimentally, there are besides microzymas, in both animals and plants, and among the micro-organisms of the circumambient air, spores, conides of fungi, of lichens, even actual cellules of ferments,[9] it is easy to understand that if these micro-organisms fall into fermentescible media they will develop in it, each according to its nature, and that various productions, moulds, diverse cellules, and at the same time vibrioniens, may appear in it.[10]

But in all the observations and experiments relative to the spontaneous change of natural vegetable and animal matters, and in the fermentations of sugar or fecula in the tissues and humors of animals, when the influence of the micro-organisms of the air has been destroyed or suppressed,[11] only microzymas and vibrioniens, and vibrios or bacteria, fruits of their evolution, are seen; this proves that the microzymas are autonomous anatomical elements existing in it of themselves.

These statements and considerations may be summed up in the following propositions:

1) The microzymas of the animal organism proceed from the vitellin microzymas, which are autonomous anatomical elements in the vitellus.

2) The number of anatomical species of microzymas is enormous.

3) The essential biological characteristic of the microzymas is that they are creators of cellules by synthesis and of vibrioniens by evolution.

4) The physiological and chemical characteristic of microzymas is to produce the zymases and to be themselves ferments having a determined form.

These propositions are also true for plants beginning with the ovule; but from the fact that a microzyma may become a vibrionien by evolution, it necessarily follows that the species of microzymas being innumerable, the species of vibrioniens are likewise innumerable.

It is important to remember that an anatomical element microzyma is animal in an animal, and vegetable in a vegetable. Hence arises this question: To what kingdom does the bacterium of a given animal microzyma belong? Or a given vegetable microzyma?

We must remember that any microzyma, before it accomplishes the evolution which produces a bacterium, passes through the evolutionary phases of microzyma slightly changed in form, of microzymas successively associated in twos, in threes, in several grains, etc.

But those forms have been described under the names of *monas, bacterium termo* and *punctum, coccus, diplococcus, torulo, streptococcus, micrococcus, mesococcus,* microbe with a point, microbe with a double point,

CHAPTER EIGHT

etc. Nor is that all; bacteria in spontaneously destroying themselves to become microzymas similar to those of the rock-chalk or of the artificial chalk of my experiments, have passed through new forms, of which the most constant is that which has also been described as the *bacterium termo*.[12]

But what are such specifications worth, being based only upon the shape, the length and thickness, the color, and the motility or immotility of the object specified? In the order of received ideas it would be too tedious to discuss them; it suffices for me to say that Dujardin, who knew the germ theory and did not allude to it in his explanations, was of the opinion that the phenomena observed in these changes were favorable to the doctrine of spontaneous generation; and consequently that outside of the microzymian theory it is all incomprehensible and arbitrary. *A priori*, one cannot tell to what kingdom a bacterium belongs, for one can only distinguish a microzyma, and consequently a bacterium, by the origin and function of the microzyma.

An example will make this clear. Take the parotid gland of a man, and that of a horse, the structure and functions of which seem to be the same and of which the microzymas of the cellules are morphologically identical. While the parotidian microzymas of man liquify and energetically saccharify the starch of fecula, those of the horse liquify that starch but do not saccharify it. And we have established by other differences of the like kind that the microzymas of the different anatomical systems of a single organism may differ one from the other; and by still greater reason those of different organisms may differ.

Plants, like animals, being anatomically determined as living by their respective microzymas, the bacteria which these microzymas can become are evidently limited to the two kingdoms; and so perhaps the question of whether a vibrionien is an animal, as was thought, or a plant, as is now asserted, is an idle one.

But if one chooses in spite of all this to insist that the bacteria are plants and that the microzymas are their spores, a new question would arise: which of the species of schizomycetes which the same microzymas may become before becoming a perfect bacterium (*bacterium termo, monas crepusculum, torula, diplococcus, streptococcus, micrococcus*, etc.)—is it first the spore, in the organism before evolution, and then in the chalk-rock or in the artificial chalk, after the total destruction of the organism?

According to accepted notions, the reply cannot be otherwise than uncertain! According to the microzymian theory, however, here is the answer:

A microzyma in a plant or an animal whose conditions of existence have just changed can become a bacterium by evolution, and the intermediate evolutionary phases, like those of the tadpole, which becomes a frog, leaves the special nature of the microzyma still existing; there are not new species. The perfect bacterium depends on the nature of the microzyma, as the perfect batrachian depends on the particular nature of its tadpole.

Every bacterium resolves itself by spontaneous destruction into a microzyma, and the microzymas thus evolved are different from the anatomical microzyma which has become a bacterium, not morphologically, nor functionally with regard to being a figured ferment, but by a collection of properties, which assure the perennity of the form and the function in a condition of individual separateness.

But the chief difference is this: the microzyma in the vitellus is the organized commencment of all animal organization, and in the ovule of the plant it is the commencement of all plant organization. On the other hand, the microzyma resulting from the destruction of a bacterium is the organized end of all organization.

And here is something stupendous! The geological microzymas, as well as those of the artificial chalk in my experiment, are organized and living, not only because without change of form, they are individually figured ferments, but also because under certain conditions, such as those of the fibrin in the experiment described in the first chapter, they act as ferments at the same time that they can again become bacteria by evolution.

The microzymas not only possess the sort of perennity of which I spoke; they enjoy also the stupendous duration of the geological epochs from the time the microzymian rocks have been formed down to the present time. This duration means that the microzymas are physiologically imperishable. And this last statement must convince us that the microzymas are organized living beings, of a class apart, and without analogue.

The following is the experimental proof that this new principle of anatomy and physiology is well founded.

The vitellin microzymas of the egg of the fowl do not pre-exist in the ovule; they are the result of a substantial development, and of the proliferation of the ovular microzymas.

To prove this, it will be sufficient to make an elementary analysis of the microzymas of the vitellus of the fowl's egg, and of those of the ovules remaining in the Graafian vesicle, while these ovules are only a few millimetres in diameter. The following are these analyses:

	Vitellin Microzymas	Ovular Microzymas
carbon	52.67%	50.63%
hydrogen	7.17%	7.36%
nitrogen, oxygen, etc.	15.71%	15.67%[13]

The difference of two percent of carbon in the percentage composition answers to great differences in the nature of the proximate principles of these microzymas. I will add that the vitellin microzymas contain much more mineral matter than the ovular. It is thus evident that the microzymas of the ovule become vitellin microzymas by substantial development, while they multiply and the vitellus grows. In short, one may say that the ovular microzymas become vitellin microzymas by maturing.

It would take too long to dwell as long as might be desirable on this result and upon the whole of the chemical, physiological and anatomical phenomena which this ripening necessitates in order that the vitellin microzymas should become fitted to play their part, chemical, physiological and histogenic, during the embryonic development. I must refer the student to what I have said elsewhere.[14]

What is most important to bear in mind is that no matter how high one goes in the scale of living beings, the microzymas are found in the ovule, and that these microzymas are not those which are to be found in the vitellus, but will become them.

All the special facts which I have made known, including the last, allow me to construct a general principle from the experimental findings; that the microzyma, the final term of anatomical analysis, is in truth the simple anatomical element which satisfies the conception of Bichat and completely destroys the idea of living matter not morphologically defined.

The cellularists, it is but fair to recall, regarding the cellule as the simplest anatomical element, believed that it proceeded necessarily from a former cellule, *omnis cellula e cellula*, holding it to be the vital unit, living *per se*, and regarded an entire organism as the sum of these units. But we now know that that was a deduction from incomplete and superficial observations, for the cellule, a transitory anatomical element, has the microzymas for its anatomical element. It is this which alone possesses all the characteristics of an anatomical element, living *per se*, and which must be regarded as the unit of life. It is what I have already stated in the following terms:

> "The microzyma is at the beginning and at the end of every living organization. It is the fundamental anatomical element whereby the cellules, the tissues, the organs—the whole organism—can be defined as living."

Let us devote a few words to develop this idea. Let us penetrate a little further into this notion of a fundamental anatomical element, which, as has been said, implies that the microzyma is the living atom of the organization as the physical atom is the element of the molecule of a simple body.

This would be true if the microzyma were unchangable in its simplicity. But in reality it is essentially mutable, as are all living bodies; and it is especially so, in order that it may fulfil its numerous functions. In fact, the microzymas, functionally different in the different anatomical systems of the same species, and different at all ages, beginning with the embryonal stage, have been primitively those of the vitellus, after having been those of the ovule.

Always anatomically simple, the microzyma becomes, by nutrition, that which it needs to become, so as to accommodate itself to each new condition of existence which the successive phases of the development of each anatomical system provide for it. It is thus that even in the embryo, in that which will be the ovary, a category of microzymas becomes again ovular microzymas to recommence the

CHAPTER EIGHT

same cycle. I add that, taken as a whole and in its details, the theory has been confirmed, verified and corroborated by a great number of other facts of general anatomy, pathological anatomy and physiology.[15]

When by the attentive study of these facts one has become convinced that the microzymian theory is their pure and simple expression, it will be at once recognized that the cell is already an organ in which, by nutrition, the conditions of the preservation of the microzymas with the constancy and regularity of their chemical and physiological functions are unceasingly realized. And it will thus be understood that the microzymas, whether of certain cells, the vitellus, or the blood, also realize after their manner the conditions of this constancy and regularity. When these conditions are no longer realized, they may undergo vibrionian evolution.

The most prominent fact in the history of the microzymas, and that which has been the most disputed precisely because of their capacity to undergo vibrionian evolution, is the fact of their anatomical autonomy. Now this faculty, which is only manifested when the normal conditions of existence of the microzymas are no longer present, is the best proof which could be given of the change which has happened in their condition, causing their irregular and changed functioning.

In fact, in their various anatomical situations, the microzymas remain morphologically true to themselves. They function in each cellule, organ, and anatomical system, naturally, chemically and physiologically for themselves, while preserving their individuality. At the same time, by coordination, they function for the benefit of the microzymian molecular granulations of the cellules, the organs and the various anatomical systems taken altogether, whose physiological condition of health is preserved by them.

But when for some reason changes occur in an organ, changes such as auscultation or percussion, Cros tells us that there can arise a discoordination, a functional perturbation in the entire organism, and disease. It is worth mentioning that from the time Dr. Cros became acquainted with the microzymian theory, he did not hesitate to recognize the microzymas as the anatomical agents of the discoordination; but how does it happen?

Among the causes which produce disease, a sudden chill in summer is the one most frequently indicated or invoked. The chill is at the same time an influence and a lowering of temperature. I do not insist on the fact that it is only something living which is painfully affected, so as to confine myself to the physical phenomenon. But the microzymas are very sensitive to variations of temperature; so much so that even the geological microzymas act regularly only at temperatures near 40° to 42°C (=104° to 107°F); in fact, the microzymas of the chalk of Sens do not cause fecula to ferment at temperatures below 38°C (=100.4°F).

Further, a very slight lowering of the temperature is sufficient for the egg which should produce a bird not to produce one, or to putrefy or produce the monsters of Dareste when heat is not uniformly applied. In fact, the influences of the medium which modify the activity of the microzymas are various. That which happens to the isolated microzyma happens also to those of the egg and those of the organism. Suppress the air and the egg does not become a fowl, but instead undergoes another kind of change.

If for any reason whatever the air does not have sufficient access to the pulmonary alveolae, and their epithelium becomes the pulmonary tubercle, the cellules become reduced to their microzymas, which are then found in vibrionian evolution in the tubercle in the cratious state. If the discoordination resulting from an irregular functioning of a part of an anatomical system is sufficient to bring on a malaise which is not removed, there will arise a diseased condition because of a sharp change in the conditions of existence of the microzymian anatomical elements, and the change in the medium sufficient to cause the discoordination will manifest itself in the vibrionian and bacterian evolution of the microzymas of the relevant part of the system.

It is thus that in the disease called *sand de rate* (anthrax), so thoroughly studied by Davaine, the diseased microzymas end by evolving into what that learned physician called *bacteridiae*, the blood gobules undergoing the changes which are so characteristic. The *bacteridiae* were not the cause of the diseased condition, but were one of its effects; proceeding from the morbid microzymas, they were capable of inducing this diseased condition in the animal whose microzymas were in a condition to receive it. Hence it is seen that the alteration of natural animal matters is spontaneous, and justifies the old aphorism so concisely expressed by Pidoux: *Diseases are born of us*

CHAPTER EIGHT

and in us.

On the other hand, the disregard of this law of nature, the firm establishment of which is accomplished in this present work, necessarily led Pasteur to deny the truth of the aphorism, and to imagine a pathogenic panspermy, as he had before conceived, *a priori*, that there was a panspermy of fermentation. That Pasteur, after having been a sponteparist, should reach such a conclusion was natural enough; he was neither physiologist nor physician, but only a chemist without any knowledge of comparative science.

What is astonishing is that he should have succeeded in procuring the acceptance of his ideas among physicians and in academies, and to procure the rejection of the microzymian theory. For instance, an enlightened physician thus summed up the fundamental proposition of Pasteur:

> "The microbes always come from without; they constitute species which recount from generation to generation up to the origin of the world.[16]

An eminent surgeon, Verneuil, ended by proposing as a demonstrated theorem that there is no spontaneous tetanus, that there is no spontaneous small pox, syphilis, glanders, hydrophobia, tuberculosis, charbon or malignant pustule; declaring that the pathogenic problem consisted solely in discovering how and when the microbe, also called a virus, came from outside the body and penetrated into the organism; declaring that the question is thus stated between old medicine and the microbian medicine "with extreme simplicity and without the least ambiguity.[17]

But these assertions of Surgeon Verneuil are reduced to nothing when we call to mind that the pretended germs of the air are only the microzymas of organisms which have disappeared, the microzymas having become bacteria by evolution; and that even at the Academy of Medicine I said—and no one ventured to contradict me—that no one had ever been able to reproduce a disease on the nosological roll by taking the pretended pathogenic microbe in normal air, but only in the diseased animal. And I add that just as with time the fibrinous microzymas lose the property of decomposing oxygenated water, so also after a short time the blood of an animal which has died of anthrax no longer communicates that diseased condition, and the same is true in all cases.

Thus normal air not only does not, but cannot, contain the pretended pathogenic microbes, and the very principle of microbian medicine constitutes a fundamental error.

But no attention was paid to this. Abandoning the famous dogma of the closure of the body to germs from without, it was held that "the human organism carries constantly a large number of microbes of many different species" which only await the moment when "the organism being disturbed in its physiological functioning will be given over to the activity of its own microbes; whose presence it had theretofore borne without being affected." Jaccond wrote this nonsense with reference to cases of acute pneumonia following a chill.[18]

In Pasteur's set, Jaccond's opinion was accepted; and although their master had declared that cellules were not living, his disciples imagined that the leukocytes (under the name of *phagocytes*) were living, like amoeba, and able to perform movements called amaeboid. And it was imagined that these phagocytes formed themselves into troops to pursue and devour the microbes. There was thus a *phagocytosis*,[19] which was trumpeted forth as providential.

A precise knowledge of the blood reduces to its just value this latest form of the struggle against the microzymian theory. Of all the suppositions and fancies of Pasteur, there remains only, even for his disciples, only a sole cause, the germs or microbes of the air, to explain the phenomena of fermentations and disease.

Nevertheless not all physicians thought as did Verneuil or Jaccond. Before 1866, while the triumph of the microbian doctrine was in full swing, Dr Tripier did not accept that there was a microbe come from without to be considered. His attention had been drawn to the new opinions by considering how frequently in the classical books of medicine a sudden chill led to everything. Here is the masterly way in which he explained it:

> "It is not at the time when the consideration of the individual coeffiecient tends to take a larger and larger place in nosological speculations that we must return to a simple etiology which has been rightly questioned. I am far from pretending that the savants to whom we are indebted for such interesting researches in the

CHAPTER EIGHT

> direction of specific causes design to bring everything within it, but those who do not exhibit that much prudence must be reminded that to constitute a morbid state, the concurrence of many conditions are indispensable, and that however specific it may be, a single cause is no cause at all."

It was thus that Tripier placed in parallel etiology according to both ancient medicine and microbian medicine. I will state later the profound meaning of the expression, drawn from algebra,[20] of "individual coefficient."

Let us say, at first, that it has been supposed that maladies resulting from specific causes are poisonings by living matters capable of reproducing themselves in the organism. The mechanism of these poisonings, says Tripier, "has been explained in many ways without being permitted to reject one on account of another."

> " According to Pasteur," he said, "the multiplication of microbes would be the consequence of the introduction of germs introduced from without. For Béchamp the microbe[21] might proceed from a special mode of evolution of living molecular granulations which he named microzymas, granulations which exist in all protoplasm, the vicious evolution whereof might be regarded as causes independent of the introduction of leaven of foreign origin."

The radical difference between the principle of microbian medicine and that of the microzymian theory of disease is thus clearly expressed. The microzymas are not the cause of disease, but by their defective or morbid functional evolution under the various influences which I have described, their evolution may become vibrionian. It was only through the ambiguity that Pasteur succeeded in creating that Tripier was able to say that I had believed that the microbe proceeded from the microzymas, and that later Jaccond thought that the microzymas are the special microbes of the human organism.[22]

To appreciate the antinomy between the microbian system and the microzymian theory, and to give to this work its practical utility by showing that the microzymian theory explains what the microbian system is powerless to make clear, it will be sufficient to recall the two fundamental facts upon which rest the fabric of the demonstration that the blood is a flowing tissue, and like all tissues, is spontaneously alterable.

The first is that a mixture of proximate principles, under the specified conditions, is naturally unalterable; but on contact with common air the same mixture always changes, owing to the various ferments which develop in it from the germs carried in this air. This mixture then does not alter spontaneously.

The second is that a natural animal matter, tissue or humor, withdrawn from a living animal in perfect health and under the same conditions, inevitably alters, even when absolutely protected from the air and its germs. Natural animal matter, then, is spontaneously alterable.

It is also desirable to recall, firstly, that the differences in the nature of the two orders of substances is such that in the alteration of the former, the micro-organisms consist of several categories of different species; while in the alteration of the latter only one category is to be found, i.e. the microzymas, and afterwards, most frequently, the vibrioniens, products of their evolution.

Secondly, corroborating the facts, creosote in adequate quantity hinders the alteration of the former in contact with a limited volume of air, preventing the appearance of ferments, while the same quantity does not hinder the alteration of the latter, nor in suitable cases, the vibrionian evolution of the microzymas.

Of these two facts Pasteur has only regarded the first and has denied the second, and it is because he and those who have trusted his word have looked upon the animal body only as organs constituted of a mixture of immediate principles —protoplasm—where nothing exists capable of becoming a vibrionien, that they have thought that the microbe coming from without is the sole cause of the alteration of this mixture and of disease.

Now if the organism was what they think, and the sole cause of disease was what they say, i.e. a mixture of immediate principles necessarily altering on exposure to the air, everyone would, of equal necessity, become diseased; but even in times of epidemics the majority are not attacked! An explanation of this fact has been sought in the microbe itself and in other considerations of the like order; but they are all worthless, for if the air contains that which changes the mixture,

CHAPTER EIGHT

it does not contain that which causes disease.

The old medicine explained immunity from disease by the receptivity or predisposition to the disease which those who are not attacked do not possess. Tripier, more precisely, invokes the individual coefficient. But a mixture of proximate principles which when exposed to the air is *always* ready to be altered enjoys no immunity!

In exact language one can speak only of the receptivity of the individual coefficient, of that which is regarded as a living body. But what is a living body? What is life?

Life, say some, is a special force, manifesting itself in ponderable matter. Mayer denies this. However it may be, they, the former, speak of a physical theory of life. We have seen that according to Pasteur, life is that which elaborates the proximate principles, the natural substances of which the organism is composed.

Bichat said: "Life is the totality of the functions which resist death." But what is life? What is death? And what is the individual coefficient in the microzymian theory? For there is no longer any question of protoplasm!

Bichat said that life was a property of tissue because he regarded elementary tissues as the living elements of organized beings, which, in his view, possessed in themselves a permanent principle of reaction which enabled them to resist the causes of destruction which surround them. The microzymian theory verifies the conception of Bichat even on this point; in fact:

The microzyma is the fundamental anatomical element, autonomically living, proliferating, while remaining morphologically similar to itself. It is in reality an apparatus whose functions manifest themselves in a medium which supplies the necessary conditions for its existence, by chemical reactions which cause it to produce the special zymases depending upon its special nature and the various proximate principles of the place and the medium in which it functions in the organism. Isolated from the organism in new conditions, as in the case of fibrin, there are some which act like lactic ferment with regard to fecula.

In short, the microzymas resist so well the ordinary causes of destruction that, in the calcareous and other rocks, geological microzymas are to be found, still living, which functioned as anatomical elements in the animals of the epoch of those rocks. Here then we have the organized being, living *per se*, physiologically imperishable, unsuspected until I described it. It is in the microzyma alone, functioning as an anatomical element, that there resides the permanent principle of reaction which enables the organisms, of which it composes the cellules, tissues, and organs, to preserve themselves by nutrition and resist the athmotelluric (Tripier) conditions which unceasingly tend to destroy them.

There is no anatomical element simpler than the microzyma, and none other like it in its resistance to total destruction. If we call life the totality of the anatomical properties which allow microzymas to construct of cellules by synthesis, and which also makes makes them capable of becoming bacteria or vibrios by evolution; and if we also define life as the aggregate of the physiological and chemical energies which enable the microzymas to produce the zymases and to nourish themselves by transforming for their own use the materials of the medium in the anatomical systems in which they function, eliminating at the same time that which they disassimilate after having used it—then it must surely be admitted that LIFE is in microzymas allied to matter, but to the matter in the *structured organization*, morphologically defined, and not simply to ponderable matter.

We now know that the microzymas are functionally different in the various anatomical systems of the same animal, and that they may be functionally different also in the same organs of the same structure in man and animals. It therefore results that it is not always permissible in experimenting to draw conclusions from one animal to another, and least of all to man. So that if we could admit with Bichat that life is a property of tissue, this property is not the same in all the tissues of the same structure and in the microzymas.

I will endeavour to explain my opinion of the reason that one kind of zymas is produced by one microzyma and another kind by another microzyma.

If there is the life of a microzyma, the life of a cellule and that of the organs of an anatomical system, there is also the life of the organized whole. This necessarily results from the coordinated entirety of the particular lives of the organs and thence of the individual lives of the microzymas which function in them. It is this view of the functions which Bichat called the entirety of the functions which resist death. But if the microzyma is physiologically imperishable, what is the death of the living individual being? It is the opposite of that which constitutes its life, i.e. it is the absolute discoordination of the functions of its microzymas.

It is thus that in a part abstracted from a living animal, i.e. muscle or blood, etc, nothing is dead; but the microzymas, the only things antonomically living, being in a state of discoordination, are no longer in their normal condition of existence. They now function only for themselves, determining the changes which attend the disorganizations of the tissues and the destruction of the cellules.

Now what is the meaning of the expression, "individual coefficient," introduced into medical language by Tripier? Just as in algebra a quantity is said to be a function of another upon which it depends, so in the microzymian theory it may be said that an organism or a cellule are quantities which are functions of the microzymas which compose it and upon which it depends. Thus the expression of coeffiecient applied to the number which multiplies these quantities can be readily understood.

The individual coefficient is the factor which increases or diminishes in the microzymas the sum of the energy which enables them to resist the various causes which, by disturbing their functioning, determine morbidity in them, and thence disease and death.

The factor, whatever it may be, being the same, and the variable, i.e. the microzymas, differing, the results will necessarily vary. Now it is a proven fact that the microzymas are functionally different between species, between races, and even in the individual, according to sex and age, and in the different anatomical systems. The individual coefficient, then, is relative to the functional differences of the microzymas of the individual.

The state of perfect health results from the constancy and regularity of the coordinated functioning of the organs within which the microzymas are healthy. But even in a state of coordination, it is necessary to take into account heredity, diatheses and atavism, which may in some way have affected the microzymas of a particular individual.

The individual coefficient, then, is a complex constant, dependent upon the particular coefficients of the relevant functional systems of the individual.

To return to the blood; here is a typical example which justifies the above considerations.

I said that in anthrax the bacteridia, instead of being the specific cause of the disease, were actually the result of the evolution of the microzymas of the blood, having become morbid as the consequence of a discoordination or of some disturbance in the physiological functioning of the organism. But it is evident that if the interior medium were inert or passive, this discoordination, in such a mixture of proximate principles, would be an effect without a cause, nothing leading it to become disturbed in its supposed functioning (for such a mixture has been shown to be unalterable of itself), while on the contrary it would immediately and infallibly be placed in a condition of alteration determined by the agent, microbe, or specific ferment introduced from outside the organism.

In short, on the hypothesis of a pure interior medium, a mixture of proximate principles, and a microbe whose multiplication is poisonous, all sheep would be equally susceptible, especially in times of epidemic, to contract anthrax under identical circumstances, by contagion, and in all cases by inoculation.

Well, this does not happen. The adult African sheep is refractory to anthrax; it does not contract the disease by contagion, and generally not even by inoculation. The individual coefficients of French and African sheep are not the same under identical circumstances. And as proof that the coefficient differs according to age, it is enough to state that the African lamb is not refractory, while the adult sheep of the same race is. Let us then say that the microzymas in the blood of the African adult sheep are among those which, even when ill treated, do not undergo that vicious alteration which causes them to become carbuncular; with the lambs of the same race it is otherwise.

If the internal medium were the mixture imagined by microbian medicine, the foregoing facts would be incapable of explanation. For the medium would be inert and passive; since it has been proved that such a mixture is always disposed to allow the multiplication of microzymas or of another like specific ferment able to alter it for its (the ferment's) own nourishment, and that the medium without the

CHAPTER EIGHT

ferment would be unalterable under other ordinary athmotelluric influences, cold, etc. It is the individual coefficient in relation to the functional differences of the microzymas of the subjects which alone explains the immunity of some, and the susceptibility of others, since it has been demonstrated that in the interior medium there is nothing autonomically living, acting and physiologically impressionable except the microzymas.

In the language of the old medicine, immunity and susceptability are determined by the capacity of the living organism to resist the influence of external or internal agents. The microzymian theory adopts this thoroughly physiological language since it is only the microzymas of the living organism which can receive impressions and either suffer or not suffer their influence, or in other words resist or not resist the perturbing causes of their functioning depending on whether the individual coefficient is abnormal or normal.

But what proof have we of this resistance, and of the mechanism of the harmlessness of the microzymas from without? The following is one such proof.

The isolated microzyma of beer yeast performs the function of a lactic ferment, producing little alcohol. In its function of an anatomic element in the globule of beer yeast, it never produces lactic acid. The young yeast, vigorous, acting strongly on cane sugar, even in contact with the air and with the addition of the chalk whose microzymas always effect lactic fermentation, still does not produce lactic acid; it resists, and microzymas of the chalk when added also fail to produce it. But if the beer yeast is old, or in some respect altered,[23] and even protected from the air, it will produce lactic acid, and the quantity will be greater if calcareous rock or even pure carbonate of lime is added.

Here we have the immunity of the beer yeast organism and its acquired susceptibility—the immunity which enabled it to resist the influence of the microzymas of the air and the chalk, annihilating their influence; and the susceptability which enabled these microzymas to produce lactic acid without hindering those of the chalk in performing their work. Here we have a picture of the immunity and susceptability of the microzymas of the cellules and tissues of the internal medium of an animal organism.[24]

In microbian medicine, the language of the old medicine is without meaning, since the former states that one sole cause produces disease and the alteration by fermentation of organic matter in general, making no distinction between the internal medium and a mixture of proximate principles.

The insuperable contradiction which exists between the microbian doctrines and the microzymian theory of the living organization brings into strong relief the truth of the aphorism of Tripier. A single cause of disease and the alteration or fermentation of proximate principles, however specific it may be, is no cause at all.

Yes, "the sole cause" is no cause, for I have demonstrated beyond dispute that there do not exist (I do not say *germs*; the word is now unsuitable) pre-existing microzymas, pathogenic or not; but there *do* exist microzymas, the living remains of bacteria derived by evolution from the anatomical microzymas of organisms which have disappeared or are disappearing before our eyes.

I limit here these considerations, referring the reader to my earlier publications, which this present work completes and corroborates.[25]

And now I hope it will be confessed that the error, common to all contemporary experimenters who have sought to discover the cause of the phenomenon of the spontaneous coagulation of the blood, and also the cause of other equally spontaneous alterations, or who, like Pasteur, maintain the natural inalterability of the blood and of all natural organic matters, is that they have regarded protoplasm as a mixture of pure proximate principles, and have held as dogma that this mixture is living and organized, although not morphologically constituted. At last I hope that it will be recognized that the discovery of the microzymas verifies the time honored conception of Bichat, according to which only that which is structured and morphologically determinate in any organism is living.

It is the agreement of the microzymian theory with the conception of Bichat which gives to the theory of the blood as a flowing tissue and to the physiological and anatomical theory of its coagulation and other spontaneous alterations their highest degree of certainty.

By way of conclusion, the following is a summary of the fundamental facts, the discovery of which has led to an understanding of the true anatomical and chemical constitution of the blood and to the explanation of its spontaneous alterations.

CHAPTER EIGHT

1) Ordinary air, near the earth, contains living microscopic objects called germs, and these germs are essentially microzymas.

2) Proximate principles, and any mixture of such principles, are unalterable in the presence of water, or of a limited volume of air at ordinary temperature when a little creosote has been first added; and such proximate principles under such conditions do not permit any organized beings to appear.

3) Natural organic matters, vegetable or animal, tissues and humors, under like experimental conditions, always change of themselves, by a phenomenon of fermentation, and at the same time the microzymas give birth to vibrioniens by evolution.

4) The fibrin of the blood is not a proximate principle; it is a false membrane containing microzymas, wherof the intermicrozymian gangue is a specialized albuminoid substance.

5) It is owing to its microzymas that fibrin decomposes oxygenated water, that it liquifies starch of amidon and that it can be dissolved, undergoing chemical change, in very dilute hydrochloric acid.

6) The microzymas of fibrin in liquified starch undergo vibrionian evolution notwithstanding the presence of creosote.

7) Fibrin liquifies spontaneously in carbolized water without the microzymas undergoing vibrionian evolution.

8) The fibrinous microzymas are special; they can produce lactic and butyric fermentation in liquified starch.

9) Natural albuminoid matters are mixtures, reducible by direct analysis into exactly defined proximate princeples.

10) The albuminoid matters reduced to proximate principles are very complex modules composed of less complex ones, amides, and their derivatives of the fatty and aromatic series. There exist of such less complex molecules, constituting an albuminoid molecule, quaternaries like urea; quinaries like taurine, which is sulphuretted, and hematosine, which is ferruginous; casein, in addition to the sulphuretted molecule, contains one which is phosphuretted; it has then six elements.

THE BLOOD AND ITS THIRD ELEMENT

11) There are several fibrins constituted as are those of the blood.

12) There are a great number of different specific albumens which coagulation does not differentiate.

13) The zymas are special albuminoid matters, likewise definable as proximate principles; they are always a functional product of the microzymas.

14) The yellow liquid of the blood, besides its albumen, contains haemozymas.

15) The haemoglobin of the red corpuscle, reduced to a definite proximate principle, decomposes oxygenated water by its noncomplex feruginous molecule, haematosine, and becomes colorless.

16) The red corpuscle of the blood is a true cellule, having a cell wall and its proper content. This content is constituted especially of haemoglobin and microzymian molecular granulations, the microzymas of which decompose oxygenated water as do those of the fibrin.

17) The blood contains a third anatomical element; the haematic microzymian molecular granulations. It is the albuminoid atmosphere of these granulations which form, by allotropic transformation, the intermicrozymian gangue of the false membrane called fibrin.

18) The blood, a flowing tissue, is the content, while the vessels, arteries, veins and their appendages are the container.

19) The three orders of anatomical elements of the blood only find their conditions of existence complete in their natural surroundings.

20) After issuing from the vessels, these conditions of existence are no longer fulfilled, and the alteration of the blood commences.

21) The microzymas of the different parts of the circulatory system possess alike the property of decomposing oxygenated water without this property being unique to them, for the microzymas of almonds, other parts of plants, and beer yeast also possess this

CHAPTER EIGHT

property. But there are animal tissues whose microzymas do not disengage the oxygen of oxygenated water.

22) The microzymas, anatomical elements, are living beings of a special order without analogue.

23) The spontaneous changes of natural animal matters, whether the microzymas have or have not undergone vibrionian evolution, thanks to free access of air, lead always under certain conditions to the complete destruction by oxidation of the product of those changes; that is to say, they are reduced to the mineral condition, carbonic acid, water, nitrogen. But the microzymas under whose influence the oxidation is effected are not attacked. After all which is purely proximate principle in a tissue, a cellule or the bacterium has undergone total destruction, the microzymas remain, and bear testimony to the existence of the vanished organization.

24) The geological microzymas of certain calcerous rocks and chalk, and those of the dust of the streets and the air, bear testmony to the microzymas which functioned as anatomical elements in the tissues of organisms of past geological epochs, even as they function in those of the present time.

25) That which in the air have been called germs are essentially the microzymas resulting from the total destruction of a living organism.

26) Normal air contains neither pre-existing germs nor the things which have been improperly termed microbes, supposed to be produced by parents resembling them.

27) The air normally contains no pathogenic microzymas. The carbon bacteridium of Davaine is the product of the evolution of diseased microzymas, either of haematic microzymian molecular granulations, or those of the blood globules.

28) There is no living matter which is not morphologically defined; that which has been called protoplasm in the cellule always contains microzymas as anatomical elements.

NOTES

1. See the memoir on *The Albuminoid Matters* pp.140 and following.
2. For the mode of multiplication of the microzymas see *Les Microzymas*, pp.490.
3. The memoir above mentioned p.162.
4. See J. Béchamp, *Normal and Pathological Albumins*, p.77 and following.
5. *C.R.* Vol. LXX p.914 (1870).
6. Conference at the Congress of the French Association for the Advancement of Science, Nantes, 1875.
7. See *Les Microzymas*, etc pp.624.
   See also note: *C.R.* Vol. LXX p.914 *Les Microzymas*, etc p.952.
8. See for details, *C.R.* Vol. LXXIV p.629; Vol. LXIII p.451; and *Les Microzymas*.
9. *Sur L'origine des ferments du vin*, by A. Béchamp *C.R.* Vol. LIX p.626 (1864).
10. See *C.R.* Vol. LXXIV p.115 and *Les Microzymas*, etc, p.948.
11. Here a complementary explanation is necessary to explain more clearly the mode of action of creosote in the experiments in which it has been employed to annihilate the influence of germs of the air. And first of all, in speaking of germs, it no longer relates to this vague *something*, which when called upon by Robin to define, Pasteur called "origin of life," but figured ferments, upon which creosote exercises an influence clearly determined. I must therefore recall that I have several times insisted on the fact that creosote is efficient in annihilating the influence of the germs of a limited volume of the surrounding air, unless the air be renewed. And it is so, because a limited volume of air contains only a limited number of micro-organic ferments. But creosote, while it does not prevent the ferments from acting, hinders their multiplication. In reality the ferments of a limited volume of air, which are capable of acting upon a fermentescible medium, do act upon it, but only in proportion to their quantity, is such a way that the result is so inappreciable that it is as though it were nothing. It is thus that the quantity of sugar, inverted, in the presence of creosote, by the microzymas of a small limited volume of air can be determined neither by reagents, nor by the polariscope. But if a slow current of several hundred litres of the same air is caused to act upon a creosoted solution of sugar the microzymas and other micro-organisms retained by the liquor render this at last cloudy, and, thus accumulated, there are among them some which effect the inversion, without developing moulds, while the microzymas undergo a greater or less vibrionian evolution. Such is the exact idea to be formed of the influence of the creosote, and of the role of the atmospheric ferments. When owing to their presence, productions such as moulds are produced, it is because the special conditions of existence of these moulds, etc., have been realized. But microzymas in their function as anatomical elements only become vibrionians from the substance of tissues and humors, even in spite of the presence of creosote, provided the volume of air is limited or completely absent.

CHAPTER EIGHT

12. See on this subject Felix Dujardin's *Les Zoophites Infusoire*, p.232.
13. See *Memoire sur les Matieres Albuminoid*, p.161 and the note on p.489.
14. See *Les Microzymas*, pp.487 and following.
15. See the following notes and publications:

    A. Béchamp: *Facts useful for the history of the origin of the bacteria.* Natural development of these little plants in the frozen parts of certain plants. *C.R.* Vol. LXVIII p.466 (1869).

    Estor: *Note for use in the history of the microzymas contained in animal cellules. C.R.* Vol. LXVIII p.519. It relates to the microzymas in bacterian evolution in a cyst which had just been removed.

    A. Béchamp and Estor: *On the microzymas of pulmonary tubercle in the cretacious state. C.R.* Vol. LXVII p.960 (1868). It relates to the discovery of microzymas in a condition of evolution within the tubercle, regarded as the remains of the destroyed epithelium of the pulmonary alveoli.

    A. Béchamp and Estor: *Facts useful for the history of the microzymas and bacteria.* Physiological transformation of bacteria into microzymas and of microzymas into bacteria in the digestive tube of the same animal. *C.R.* Vol LXXVI p.1143 (1873).

    A. Béchamp: Facts useful for the history of the histological construction of the glairine of Moltig. *C.R.* Vol. LXXVI p.1485 (1873).

    A. Béchamp: The diseases of the silk worm. *C.R.* various notes from 1866 to 1874. They relate to the pebrine, a parasitic disease, and to the flacherie, a microzymian disease, not parasitic.

    J. Grasset: *On the histological phenomena of inflammation.* Treatise regarding a new theory, based upon the consideration of the molecular granulations (microzymas). Gazette Med. de Paris, year 1873.

    E. Baltus: *Theory of the Microzyma*, a theoretic and practical study of pyogenesis (the formation of pus). Theses of the Faculty of Montpellier, year 1872. No. 41.

    J. Béchamp: *The Microzymas and their functions at the different ages of the same being.* Theses of the Faculty of Montpellier, 1875 No. 63.

    A. Béchamp: *Microzymas and disease*; in *Les Microzymas*, etc p.744. (Chamalet, 60 Passage Choiseul.)

    A. Béchamp: *Puerperal septicaemia, pleurisy, the albuminuria* and the preface to *Microzymas et Microbes*. (Chamalet, 60, Passage Choiseul, Paris.)

    A. Tripier: *Electricity and Cholera: Genesis, prophylaxy and treatment.* (Georges Carre, pub. 1884). In this memoir there will be found a comparison of the microbian system and the microzymian theory, highly original and at the same time the conception of what the eminent author terms the *individual coefficient*.

16. *Gazette medical*, Paris 6th Series, Vol. V, p.218. This is precisely what Chamberland said of micro-organisms in general: *Recherches sur l'origine et le development des organismes microscopiques.* Theses de la faculte des Sciences, Paris, 1879. See also *Microzymas et Microbes*, p.25, 2d ed.

17. *C.R.* Vol. CV p.552.

[There is an implication to be found in the statement of Surgeon Verneuil, though probably not meant by him, to which assent must be given when understood. It is *true* that there is no such *thing* as tetanus, small pox, syphilis, etc, as is implied by the general use of nosological terms. Disease is not a *thing*, nor an *enitity*; it is a *condition*, and the error of regarding the condition of disease as an entity has confirmed, where it has not originated, much of the prevailing erroneous treatment of the sick.

Nosological terms have a use; it is that of bringing to the mind of the physician a group of pathological symptoms, which may or may not be present in the case of the patient under consideration; from them, when present, the diseased condition of the patient can be recognized and treated. Unfortunately, through not understanding this truth, attempts are frequently made to treat, not the patient, but the name, which has been given to a collection of morbid symptoms.—Trans.]

18. *Journal des societes scientifiques*, 4 May 1887, p.156.

19. [These words must be erased from the language of science.—Trans.]

20. [The term *individual coefficient* was first introduced to indicate the amount and direction of errors which each individual astronomer was prone to commit.—Trans.]

21. [The term *microbe*, introduced for the purpose of drowning the grand discoveries of Béchamp, is, as presently shown, an etymological solecism.—Trans.]

22. This is how the ambiguity was created. The surgeon, Sedillot, thoughtlessly invented the word *microbe* as a name for the vibrioniens, which eventually Davaine regarded as the living agents of disease. Pasteur, heedless of the inaccuracy, (even in the etymological sense of this word being applied to a microscopic being of immense longevity) adopted it to designate the micro-organized ferments; thus beer-yeast was a microbe, as also the bacteridia of Davaine. He went further, and in a book published under his auspices he permitted the following definition to appear: "Under the name of microscopic beings or microbes are meant all living beings too small to be seen by the naked eye, all those which can only be seen with the aid of instruments which can enlarge them a great number of times, such as the small worm called trichina, which produces trichinosis, and an acarus, which produces the itch..." The work from which the above is quoted appeared in 1883 with a preface by Pasteur. Here we see how all diseases are regarded as parasitic on the same ground as the itch, and how microzymas have come to be miscalled microbes!

23. [The French text is *aleree*, which, I believe, to be a press error for *alteree*,—Trans.]

24. [We are here presented with a forcible illustration of the reckless ignorance of those physicians who practice the inoculation of organic poisons, such as the products of diseased conditions known as vaccines, anti-toxins, etc, upon man and other animals, whether as preventives or remedies. Even the changes mentioned in the text, as some of the results of the experiments of Professor Béchamp, are unknown to these gentlemen; and, absolutely ignorant of what effect such inoculations may have upon the life forces, i.e. the microzymas, of their victims, they arrogantly insist that their ignorance is learning, and induce a degeneration among those races who, recognizing their ignorance, place their faith in men as ignorant as themselves!—Trans.]

25. For general pathology, see the three last conferences of *Les Microzymas*. For special pathology, the communications *Sur la Septicaemic Puerperale, Sur la Pleuresie* and *Sur les Albuminuries*, in *Microzymas et Microbes*. And for the physiological theory of fermentation, as well as for the true theory of nutrition, various chapters of the same works. (Chamalet, publisher, 60 Passage Choiseul, Paris.)

POSTFACE

This postface consists of a note read before the Academy of Medicine on the 3rd of May, 1870. It establishes an important date in the history of science during the last three decades of the last century. The microbian doctrines were not yet imagined; nor were they, till several years after, as a result of the plagiarizing of the microzymian theory.

The microzymas, pathology and therapeutics.[1]

Chauffard has recently published an important work on the treatment of smallpox by carbolic acid. His conclusions interest me greatly, and I desired to make the matter clear to the Academy.

In a note which appeared in the *Transactions of the Academy of Sciences* (Vol LXVI, p.366), I said, in reference to a note of Chauveau on the molecular granulations of the vaccine virus:

> "The transition from the study and meaning of the molecular granulations which are born or act in certain fermentations and which I have named *microzymas*, to the study and meaning of those which exist normally in all the tissues of organized beings, and also in the cellules of those tissues, was natural.
>
> My satisfaction, then, was extreme when I saw Chauveau enter upon this path, and, from another point of view, confirm the observations made in the laboratory of the chemist. I said "from another point of view"; I was wrong, because from the physiological point of view where I had placed myself and whence I studied what is called fermentation, the experiments of Chauveau, on the molecular granulations of the vaccine virus, are closely connected with mine.
>
> I place the molecular granulations in solutions of simple organic matters; Chauveau in the organic and organized matters of living beings."

POSTFACE

From a time long ago certain diseases have been compared to fermentations. We may go back to Stahl and Willis and probably still earlier for this, though that is not important, for, as was remarked by Babinet, "Antiquity has told everything; when it told truly, it was simply a wonderful accident, and it proved nothing."

My researches upon fermentations and ferments, particularly upon molecular granulations, date back some fifteen years, and those which Professor Estor and I conducted for the purpose of generalizing my earlier observations have led to this result: that the animal is reducible to the microzyma. But the microzyma, whatever its origin, is a ferment; it is organized, it is living, capable of multiplying, of becoming diseased and of communicating disease.

All microzymas are ferments of the same order—that is to say, they are organisms, able to produce alcohol, acetic acid, lactic acid and butyric acid.

In a state of health the microzymas of the organism act harmoniously, and our life is, in every meaning of the word, a regular fermentation. In a state of disease, the microzymas do not act harmoniously, and the fermentation is disturbed; the microzymas have either changed their function or are placed in an abnormal situation by some modification of the medium. This was what I tried to make clear by a positive example of a kind which would leave no room for misunderstanding either the extent or the bearings of the conclusion.

The harmonious function of a bird's egg is to produce a bird. During incubation the chemical acts which are accomplished within it result in the transformation of the materials of the yolk and the white into the various chemical compounds which will form the various organs of the complete animal.

While these chemical acts are being accomplished, no gases other than the normal gases of respiration are set free. But, if that which will be the embryo is abstracted from the egg, it contains nothing organized but the microzymas. That which will be the embryo is itself, at first, only a collection of microzymas. From the chemical point of view, everything within the egg is the work of the microzymas.

What will happen if in the egg we proceed to mix up those elements within the egg which were not destined to be mixed together? Donne said and demonstrated that the egg becomes putrid. I am of the same opinion, but this has to be explained.

If, as was done by Donne, everything in the egg is mixed up by violent shaking, there is soon observed an escape of carbonic acid gas, hydrogen and a trace of sulphuretted hydrogen. When the escape of gas has ceased, the contents of the egg, from being alkaline as it was before the mixture, have become acid; the odor is disagreeable, but gamey only, distinct from the horrible odor of rotten eggs, which are alkaline.

If we then examine what has happened to the materials of the egg, the albuminoid substances and fatty matters are found to be unchanged. The sugar and glucogenic matters have disappeared, and in their place we find alcohol, acetic acid and butyric acid. What has then taken place has not been a putrefaction, but a distinctly characterized fermentation. The violent agitation has not killed anything which was organized within the egg; only the order of its contents has been disturbed.

The microzymas have been thrown into media which was not intended for them; those of the white into the yolk, and vice versa. Having been forced to take their nourishment from materials not intended for them, they have reacted in a new manner, but without any change in their nature or appearance.

I could multiply such examples and show that the same microzyma, free or enclosed in a cellule, acts in the former condition as a lactic or butyric ferment, in the latter as an alcoholic ferment. I have reported the example of what happens in the egg because in this instance nothing foreign intervenes; fundamentally, the egg is an animal *in posse*.

But the microzymas may be regarded from another point of view. Not only are they individually ferments, but they are also able to produce bacteria.

This ability, alike for all, does not manifest itself equally for all under the same conditions. This amounts to saying that in each natural group of beings, and also within each centre of activity within each organism, the microzymas possess a certain specificity.

What I mean is that the microzymas of dogs, sheep, birds, etc, and those of the liver, the pancreas, or the blood, for instance, although morphologically identical in appearance, and even identical in certain aspects chemically, are nevertheless different. What is remarkable is

POSTFACE

that the bacterium derived from the microzyma possesses the same function as that microzyma; it is a ferment of the same order.

Not only is the microzyma a builder of the bacterium, but it is also a builder of the cellule; but in this new condition its functions may be entirely changed. The microzymas which are butyric ferment, and which produce bacteria which are also butyric ferments, may produce cellules which are alcoholic ferments.

Finally, the microzyma may become diseased and may communicate the diseased condition.

The first time that my attention was called to this subject was in relation to my studies of the diseases of the silk worm. On examining the eggs of a nursery in which there were many morts-flats, I was struck with the presence in these eggs of molecular granulations, motile like the others, but more abundant, of which a large number seemed united in 2, 3, and 4 grains, like the chaplets of microzymas.

I asked myself if there might not be a relation of cause and effect. All the eggs which presented this characteristic yielded morts-flats, and those worms which did not die produced butterflies which in turn produced eggs possessing this same character. Finally, when the disease was at its worst, the animal and sometimes the eggs contained bacteria. There is then for the silk-worm a characteristic which enables one to say, *ab ovo*, that the caterpillar which will be born of this egg will be afflicted with a certain disease.

I have not yet had an opportunity to study the different viruses from this point of view, but there can be little doubt that those of smallpox and syphilis contain specific microzymas, i.e. they transport the disease of the individual from which they originate. These two examples have led to the proposal of the specificity of certain diseases called infectious. I do not contradict this. Nevertheless, when we see that smallpox and syphilis are inoculable upon certain animals, and that anthrax is not communicable to dogs nor yet to birds, it is certainly right to ask why.[2]

Notwithstanding many remarkable works, nothing is more obscure than the cause which presides over the development of diseases and their communicability. But what we can affirm is that when we are sick, it is we who suffer, and that the suffering is a cruel reality. This is because the cause of our diseased condition is always within ourselves.

External causes contribute to the development of the affliction and hence of the disease only because they have brought about some material modification of the medium in which live the ultimate particles of the organized matter which constitutes us, namely, the microzymas.

These external causes, by a succession of changes brought about, and depending on a crowd of variables, bring about correlatively a further change, which then bears precisely upon the physiological and chemical status of the microzymas.

The tendency of the most recent researches is to show that miasms, like viruses, contain living microscopic organisms, something analogous to microzymas and bacteria, which proliferate in the blood or tissues of the animal and make it sick. I do not believe that things happen in that manner.

Every phenomenon having a cause, I admit the existence of organized particles in miasms, but I do not believe in their proliferation in the organism, a proliferation which has nowhere been proven, up to the present time, and which many experiments positively contradict.

Two authors, for instance, who agree in regarding the malignant pustule as a fermentation and who also agree that the blood of an animal attacked by a disease can communicate it to another animal of the same species, agree no longer when they endeavor to explain what they observe. For Davaine, the virulence of the carbuncular blood is due to the species of bacterium to which he gave the name of *bacteridium*. For Sanson, this virulence is due to a specific putrid change in the blood. According to him, the bacteria have nothing to do with it. Often they are not to be found in it; nothing organized can even be seen. He even doubts that the bacteria are animals, or plants, or even living beings. And the author remarks—and this time truly—that putrid albuminoid matter, although containing bacteria and even bacteridia, cannot communicate anthrax, even to an animal susceptible to it.

What does all this mean—if neither bacteria nor the products of the putrefaction of albuminoid matters communicate anthrax?

I will try to explain these contradictions.

Davaine made an experiment which I regard as a very important one upon this question. He inoculated some very parenchymatous plants with some putrid matter of plants, in which *bacterium termo* or

POSTFACE

something similar to it was present.

In an opuntia and in an aloe, he said, the bacteria propagated while preserving their primitive characters. Inoculated from these plants upon another aloe, they gave birth to long filaments divided into 2, 3, or 4 articles or segments. These long filaments, being innoculated upon a new species of aloe, produced corpuscles in a fine powder. Lastly the long bacteria, inoculated upon the species of opuntia and of aloe, which were the subjects of the first inoculations, reproduced the *bacterium termo*.

These facts cannot be disputed. The authority of Davaine guarantees them, but their interpretation seems to me to be open to question.

On the other hand, when I examined the frozen parts of several species of plants (belonging to various families), in which previous to the congellation there had been no lesion whatever, I always found bacteria of several kinds, not to say species, according to the specific nature of the frozen plant, and in the healthy parts, adjacent to the latter, there was not a trace of bacteria; nothing but normal microzymas. This proves that bacteria can develop in plants without inoculation, just as they can develop and even exist normally in man throughout the entire length of the digestive tract.

I would then explain the experiment of Davaine by saying that by the wound and the introduction into this wound of certain bacteria and of the liquids which saturate them, this savant produced a lesion and a change of medium which permitted the normal microzymas of the inoculated plants to evolve according to their own individual aptitudes,[3] and there was no proliferation of the inoculated bacterium.

It is the same with animals. It is not the inoculated organisms which multiply, but their presence and the liquid which saturates them causes a change in the surrounding medium which enables the normal microzymas of the organism to evolve in a diseased manner, either reaching or not reaching the state of a bacterium. The disease is not the consequence of the new mode of being of the normal microzymas; the fever which ensues is only the result of this new method of functioning and of the effort of the organism to rid itself of the products of an abnormal fermentation and disassimilation, while inducing a return of the diseased microzymas to the physiological condition.

THE BLOOD AND ITS THIRD ELEMENT

This theory, which is founded upon facts ascertained by indisputable experiments, explains, among other things, why the blood of carbuncular sheep containing bacteridia inoculated upon dogs or birds does not induce the appearance of bacteridia and the development of the carbuncular disease, as Davaine has shown.

But is there any difference in the purely chemical materials of the blood of a dog, a bird and a sheep? They contain the same albuminoid and other matters, the same salts, the same fatty bodies, and under other conditions, the microzymas found there certainly evolve into bacteria. The only difference which exists, as is proved by the experiment itself, must be in the histological elements of the blood of these animals and in their unequal susceptability. If then the bacteridae inoculated upon birds and dogs do not multiply as they should have done, it is certainly not that the chemical medium is different; and if anthrax does not result from the inoculation, it is because the microzymas of these animals are unfit to evolve morbidly under the influence of the medium which promotes the introduction of morbid materials.

To sum up, the microzymas are organized ferments, and they can under favorable circumstances produce bacteria. Under other circumstances they become builders of cellules. All organisms, *ab ovo*, are created by them. In short, the cellule, the bacterium itself, can rebecome[4] a microzyma, and thus the microzymas are seen to be the beginning and end of all organization. If that is true, we ought to encounter them wherever organized beings have lived; and the fact is that I have found them in all the calcareous rocks from the oolithic to the most recent; the dusts of our streets swarm with them, and there as everywhere they are ferments of the same order. Not all of them are morbid. If they were, we would be living under a constant menace; but there may be morbid ones among them.

What relation is there between the above and the work which I recalled at the commencement? The relation is the following:

It is now a long time, from the beginning of my researches upon ferments at a time when nobody occupied themselves with the question of spontaneous generation, since I demonstrated (in opposition to generally received ideas) that creosote (and phenic acid, for at this time this acid was sold as creosote, especially in France), in a non-coagulating dose, did not impede a fermentation that had already commenced.

I showed that in the same dose, these agents prevented the appearance of organized ferments in the most fermentiscible mixtures. And I gave as explanation for this the fact that they opposed the germination or hatching of the germs of microphytes or microzair ferments which the air might bring to the mixtures, thus confirming an old experiment of Humer, recalled by Chevreul, and precisely proved by that savant, i.e. that the vapors of the essence of turpentine, in a confined space, hindered the germination of seeds and caused the destruction of those which had begun to germinate. I also demonstrated that the same doses of these agents did not hinder fresh muscle from acting on fecula starch, to liquify and to cause it to ferment, nor finally the appearance of bacteria in the mixture.

I concluded from this that the muscle must contain ferments already developed, living, and active of themselves, since creosote did not prevent the fermentation from beginning. This observation was the point of departure for the researches that Estor and I undertook upon the evolution of the microzymas of the higher organisms into bacteria. In ending my earliest observations, in 1866, I advised the use of creosote and phenic acid in the rearing of silk worms, for the purpose of preventing the birth of the vibrating corpuscle, which is the vegetable parasite of the pebrine.

At the same period Dr. Masse, starting from the same point of view, employed the same agent to dry up the fecundity of the spores of the microsporon mentagrophytes of parasitic sycosis.

In 1868 my friend, Dr. Pecholier, inspired by similar ideas, published his researches regarding the treatment of typhoid fever by creosote; he proposed to prevent the appearance and multiplication of the typhoid ferment. Later Gaube published a work in confirmation of that of Dr. Pecholier. The same year Calvert reported the experiments made at Mauritius by Drs. Barrault and Jessier on the application of phenic acid in the treatment of typhoid and intermittent fevers.

The above is the connection of the ideas and origin of the employment of creosote and carbolic acid in therapeutics. The theory of this employment is as follows:

Creosote dries up the fecundity of the germs which produce disease,[5] in conformity with the principles enunciated by me in 1857. The following experiment, while maintaining the principle, gives it a wider meaning and places it in connection with the first parts of this discourse.

Beer yeast is a complete organism, though reduced to the state of a simple cellule. As an alcoholic ferment, in a sugared medium it preserves indefinitely its cellular form. But under other conditions, things happen differently. Beer yeast, it has been said, does not cause starch to ferment; that is an error; it causes it to ferment, but in a different manner to sugar, that is all. If it is introduced into starch of fecula, with some very pure calcic-carbonate (not from the calcareous rocks), the whole being creosoted to hinder the influence of germs of the air, the starch will be liquified, a fermentation will be set up and the yeast disappears by degrees and is finally replaced by an innumerable quantity of superb bacteria. The fermentation is acetic, lactic and butyric instead of being alcoholic.

It may be said that it was the bacteria which were the ferments; granted, but observe that these bacteria are the issue of the beer yeast, of its microzymas. That settled, in other experiments, the same quantities of yeast, calcic carbonate and starch being employed, and double and triple the quantity of creosote, the starch was still liquified, and the fermentation proceeded, but the globules of yeast were not destroyed, and the bacteria did not appear. The yeast was not killed; the creosote when used in greater amounts has only prevented the evolution of the microzymas into bacteria.

Creosote, which resists the blossoming of the germs of microphytes and microzoairs in fermentiscible media, preventing thus the commencement of fermentation, does not hinder a fermentation already commenced and where there exist already adult organisms. But in certain doses it is a moderating agent which, according to the experiments just mentioned, regulates the function of the cellule and its microzymas, which it prevents evolving into bacteria.

The explanation of the role of carbolic acid and creosote in therapeutics is easy to understand, if account is taken of the researches which have permitted this hasty resume to be made. These agents do not hinder the physiological functioning of the histological elements of the organism, but they arrest the morbid evolution of the microzymas, the too rapid destruction of the cellules, and tend, doubtless by modifying the medium, to bring back into harmony the functioning of the deviated microzymas.

This recalls unavoidably the agents used in the old therapeutic devices which our ancestors employed; camphor, essences, musk, etc. It is true that it was empirically that they fulfilled the indications which, after many deviations, we now supply like them, but instead we use new methods which rely on experimental and positive data.[6]

And, in conclusion, I beg the permission of the Academy to repeat here something which Professor Estor and I said in a recent work upon this subject:

> " After death (leaving here the domain of pathology to enter into that of the physiology of the species), it is essential that matter be restored to its primitive condition, for it has only been lent for a time to the living organized being. In recent years an extravagant role has been assigned to the airborn germs; the air may bring them, it is true, but it is not necessary that it should do so."

The microzymas, whether in the state of bacteria or not, are sufficient to assure by putrefaction the circulation of matter.

The living being, filled with microzymas, carries in itself the elements essential for life, disease, death and destruction. And that this variety in results may not too much surprise us, the processes are the same. Our cellules, it is a matter of constant observation, are being continually destroyed by means of a fermentation very analogous to that which follows death. Penetrating into the heart of these phenomena we might really say, were it not for the offensiveness of the expression, that we are constantly rotting.

THE BLOOD AND ITS THIRD ELEMENT

NOTES

1. See note at end of this postface.
2. (from p.208) The admirable caution of this true man of science is worthy of notice. When almost the whole scientific world had gone crazy over belief in the "specificity of disease," Béchamp says merely: "I do not contradict this." The translator is of the opinion that disease is not an entity—*a thing*—but a condition, and that the opinion that it is an entity is responsible for many of the errors of modern medicine. The reasons for this opinion will be given on a fitting opportunity. A large sect, calling themselves Christian Scientists, deny the reality of sickness and say that it is "an error." There can be no question that many illnesses, especially of many wealthy persons, is imaginary only, and these can be cured by Christian Science, mental healing, hypnotic suggestion and the like. I have been unable to find any rational foundation for the rest of the claims of these sectarians. The desire of so many physicians to prevent such persons from attempting to heal those who are willing to be treated by them is similar to the persecutions initiated by Torquenmada and practiced in these days by the Russian Church.—Trans.
3. [*Individual aptitudes*, that is, in the altered medium, abnormal and therefore diseased, and productive of a diseased condition, but not neccessarily that of the inoculated matter.—Trans.]
4. [Would it not be more agreeable to the facts discovered by Prof. Béchamp to regard the much smaller microzymas which result from the final evolution as the actual offspring of the parent microzymas via the bacteria, like the butterfly from its parent butterfly via the chrysalis?—Trans.]
5. [Diseased conditions—Trans.]
6. [This expression, and some statements in the parts of this chapter immediately preceeding it, have been cited by microbiologists to support the assertion that Béchamp believed in the germ theory of disease. Such a statement illustrates a consciousness of the weakness of their position and their eagerness for calling in aid occasional expressions of their opponents. The truth is that the word "germ" is used in a fast and loose way by the microbists, and there is a meaning in it which it might be said that "germs" have produced this or that diseased condition. All which serves to show the importance of exactness in the use of language, a fact rarely borne in mind or carried into practice by these savants. It would be out of place to enter into a discussion of this here; besides, the theory itself is destined to fall, "with a great bursting of bubbles," as soon as the writings of Professor Béchamp become widely known. His *Microzymas et Microbes*, and his designation of the microbian theory in *Les Grands Problems Medicaux*. p.11, as *la plus grande scottise scientifique de ce temps*, sufficiently indicate his opinion.—Trans.]
7. Note to p.205.

Here for persons whom it may interest, follows a list of memoirs and articles in which may be found the historical succession of the ideas which have enabled the resume contained in this chapter to be written:

POSTFACE

On the influence which pure water or water charged with various salts exercise at a low temperature upon cane sugar (moulds and spontaneous generations). Annales de chime et de physique, 3rd series, Vol. XLVIII (1855 and 1856) C.R. Vol. XL, p.436, and Vol. XLVI, p.44 and Annales de chimie et de Physique, 3rd series, Vol. LIV, p.28 (1858).

Memoir upon generations called spontaneous and upon ferments. Annales de Societe Linne de Maine et Loire, Vol. VI (1863), and see C.R. Vol. LVII, p.958.

Note upon alcoholic fermentation. C.R. de l'Academie des Sciences, Vol. LVIII, p.601 (1864), and Montpellier Medical, Vol. XII.

Upon alcoholic fermentation, Reply to Berthelot, C.R., Vol. LVIII, p.1116 (1864).

On some new soluble ferments (Anthozymas). C.R. Vol. LIX, p.496 (1864).

On the origin of the ferments of wine. C.R. Vol. LIX p.626 (1864).

On the escape of heat as a product of alcoholic fermentation. C.R. Vol. LX p.241 (1865).

Memoir upon nefozymase. Montp. Med., Vols XIV and XV.

On the cause which matures wines. C.R. Vol. LXI, p.408 (1865) and Vol. LXIX p.892 (1869).

On physiological exhaustion and on the vitality of beer yeast. C.R. Vol. LXI p.689 (1865).

On the harmlessness of the vapors of creosote in the breeding of the silk worm. C.R. Vol. LXII p.1341 (1866).

On the parasitic disease of the silk worm. C.R. Vol. LXIII pp 311, 341, 391, 425, 552, 693, 1147 (1866); Vol. LXIV pp 231, 873, 980, 1042, 1043, 1185, (1867); Vol. LXV p.42 Vol. LXVI p.1160 (1868); Vol. LXVII pp 102, 443 (1868); Vol. LXIX p.159 (1869).

On the role of the calcareous earths in butyric and lactic fermentations, and of the living organisms which they contain (microzymas). C.R. Vol. LXIII p.451 (1866).

Microzymas in the waters of Vergeze. C.R. Vol. LXIII p.559 and Bull. Soc., Vol. VI p.9 and Vol. VII p.159 (1866).

On the role of the microscopic organisms of the mouth in digestion, and especially in the formation of the salivary diastase; in common with Prof. Estor and Saintpierre. Mont. Med. Vol. XIX.

On the molecular granulations of fermentations and of the tissues of animals (microzymas). C.R. Vol. LXVI. pp 366, 1382 (1868).

On the nature and function of the microzymas of the liver; jointly with Prof. Estor. C.R. Vol. LXVI p.421 (1868).

On the origin and development of the bacteria; jointly with Prof. Estor. C.R. Vol. LXVI p.859 (1868).

On the reduction of nitrates and sulphates in certain fermentations. C.R. Vol. LXVI p.547 (1868).

On the spontaneous alcoholic and acetic fermentation of eggs. C.R. Vol. LXVII p.523 (1868).

On the microzymas of pulmonary tubercle in the cretacious state. Jointly with Prof. Estor C.R. Vol. LXVII p.960 (1868).

Facts to serve for the history of the origin of bacteria; natural development of these little plants in the frozen parts of several plants. C.R. Vol. LXVIII p.466; Mont. Med. Vol. XXII p.320 (1869).

Conclusions relating to the nature of the mother of vinegar and the microzymas in general. C.R. Vol. LXVIII p.877; Gazette Medicale de Paris, 8 May 1869.

On the alcoholic fermentation by the microzymas of the liver. C.R. Vol. LXVIII p.1567 (1869).

Researches relating to the microzymas of the blood and the nature of fibrin. Jointly with Prof. Estor C.R. Vol. LXIX p.713 (1869).

Note for use in the history of the microzymas contained in animal cellules. by Prof Estor C.R. Vol. LXVII p.529.

On the nature and origin of the blood globules. Jointly with Prof. Estor C.R. Vol. LXX p.265 (1870).

On the geological microzymas of diverse sources. C.R. Vol. LXX p.914 (1870).

On the carbonic and alcoholic fermentations of sodic acetate and of ammonium oxalate C.R. Vol. LXX p.69 (1870).

See also

On the circulation of carbon in nature and the instruments of this circulation; exposition of a chemical theory of the life of the organized cellule by A. Béchamp, Paris, Asselin; Montepellier, Seguin.

Of the microzymas of the higher organisms by Messrs. Béchamp and Estor. Mont. Med., Vol. XXIV p.32.

Exposition of the physiological theory of fermentation, according to the researches of Prof. Béchamp by Estor. Messager du Midi (1865).

[The student is to understand that the above list comprises but a small fraction of the scientific labors of the late Professor Béchamp. A fuller list, though still imperfect, occupies eight of the large folio pages of the *Moniteur Scientifique* (Paris) for December, 1908, and these labors were spread over fifty-three years, from 1853 to 1905 inclusive. Genius has been defined as, in one aspect at least, the "faculty for taking infinite pains," and this faculty was possessed by Béchamp in an almost infinite degree. The world has yet to learn how much it owes to this remarkable genius. The acknowledgment will be resisted by all those interests which fatten upon the ignorance and trusting confidence of the people. But thanks to his researches and discoveries it cannot be long before the medical profession will recognise the dangerous errors into which it has been led by those who succeeded in establishing a conspiracy of silence around Béchamp and his discoveries.—Trans.]

Dr Leverson wrote the above paragraph in 1912.

ALSO FROM A DISTANT MIRROR

Béchamp or Pasteur?
A LOST CHAPTER IN THE HISTORY OF BIOLOGY

ETHEL D. HUME
WITH A PREFACE BY R. PEARSON

HARDBACK
PAPERBACK
KINDLE
EPUB
PDF

"An amazing alternative interpretation of biochemical history... a compelling account of Pasteur's plagiarism and a strong reminder of the powers at work in the pharmaceutical and regulatory industry."

"For those physicians who can think outside the box of their medical training, Hume's book will challenge your thinking on 'disease theory' that has dominated American medical schools for the past century.

It is well researched and presents documentation from the Academy of Physicians of which both Pasteur and Bechamp were members. She presents excerpts of presentations which both men gave before the academy.

One sees easily the difference in intelligence and integrity between these two men."

A DISTANT MIRROR

ALSO FROM A DISTANT MIRROR

The Blood and its Third Element
Antoine Bechamp

Bechamp or Pasteur?
Ethel Hume

Reconstruction by Way of the Soil
Guy Wrench

The Soil and Health
Albert Howard

The Wheel of Health
Guy Wrench

The Soul of the Ape & My Friends the Baboons
Eugene Marais

The Soul of the White Ant
Eugene Marais

Earthworm
George Oliver

My Inventions
Nikola Tesla

The Problem of Increasing Human Energy
Nikola Tesla

Response in the Living and Non-living
Jagadish Bose

WWW.ADISTANTMIRROR.COM.AU

Printed in Great Britain
by Amazon